Elena Pancera

Strategies for Time Domain Characterization of UWB Components and Systems

Karlsruher Forschungsberichte
aus dem Institut für Hochfrequenztechnik und Elektronik

Herausgeber: Prof. Dr.-Ing. Thomas Zwick

Band 57

Strategies for Time Domain Characterization of UWB Components and Systems

by
Elena Pancera

universitätsverlag karlsruhe

Dissertation, Universität Karlsruhe (TH)
Fakultät für Elektrotechnik und Informationstechnik, 2009

Impressum

Universitätsverlag Karlsruhe
c/o Universitätsbibliothek
Straße am Forum 2
D-76131 Karlsruhe
www.uvka.de

Universitätsverlag Karlsruhe 2009
Print on Demand

ISSN: 1868-4696
ISBN: 978-3-86644-417-1

Preface of the editor

The permanently increasing demand for higher data rates in wireless communication systems imperatively leads to a continuous search for new wireless standards which enable a better use of the existing frequency spectrum. A promising new option is the ultra-wideband technology. Signals will be spread over a very wide frequency range which leads to a very low spectral power density and therewith a low interference potential for all other wireless services. Recently released frequency band allocations by the authorities in several countries worldwide paved the way for future ultra-wideband applications. However the plurality of research and development results worldwide clearly demonstrate a much higher potential for sensors using the newly available spectrum compared to communication devices.

The doctoral thesis on ultra-wideband technology of Ms. Dr. Elena Pancera is based on the finding that for such extremely wideband systems a frequency domain analysis is absolutely insufficient. Therefore she developed a methodology for a time domain system analysis and demonstrated it based on the thereto most critical components such as antennas and filters. In the second part of her work Ms. Pancera transferred the well known calibration procedure for the measurement of radar cross sections into the time domain. This very innovative work demonstrated that in case of very broadband measurements the new methodology has advantages regarding the required effort in measurement and computation.

The presented work of Ms. Pancera is a fundamental basis for future research in the field of ultra-wideband technology. The results will draw attention worldwide and entail more relevant research. I wish her all the best for her future and I am sure that she will produce many more essential contributions in science.

Prof. Dr.-Ing. Thomas Zwick
- director of the IHE -

Forschungsberichte aus dem
Institut für Höchstfrequenztechnik und Elektronik (IHE)
der Universität Karlsruhe (TH) (ISSN 0942-2935)

Herausgeber: Prof. Dr.-Ing. Dr. h.c. Dr.-Ing. E.h. Werner Wiesbeck

Band 1 Daniel Kähny
Modellierung und meßtechnische Verifikation polarimetrischer, mono- und bistatischer Radarsignaturen und deren Klassifizierung (1992)

Band 2 Eberhardt Heidrich
Theoretische und experimentelle Charakterisierung der polarimetrischen Strahlungs- und Streueigenschaften von Antennen (1992)

Band 3 Thomas Kürner
Charakterisierung digitaler Funksysteme mit einem breitbandigen Wellenausbreitungsmodell (1993)

Band 4 Jürgen Kehrbeck
Mikrowellen-Doppler-Sensor zur Geschwindigkeits- und Wegmessung - System-Modellierung und Verifikation (1993)

Band 5 Christian Bornkessel
Analyse und Optimierung der elektrodynamischen Eigenschaften von EMV-Absorberkammern durch numerische Feldberechnung (1994)

Band 6 Rainer Speck
Hochempfindliche Impedanzmessungen an Supraleiter / Festelektrolyt-Kontakten (1994)

Band 7 Edward Pillai
Derivation of Equivalent Circuits for Multilayer PCB and Chip Package Discontinuities Using Full Wave Models (1995)

Band 8 Dieter J. Cichon
Strahlenoptische Modellierung der Wellenausbreitung in urbanen Mikro- und Pikofunkzellen (1994)

Band 9 Gerd Gottwald
Numerische Analyse konformer Streifenleitungsantennen in mehrlagigen Zylindern mittels der Spektralbereichsmethode (1995)

**Forschungsberichte aus dem
Institut für Höchstfrequenztechnik und Elektronik (IHE)
der Universität Karlsruhe (TH) (ISSN 0942-2935)**

Forschungsberichte aus dem
Institut für Höchstfrequenztechnik und Elektronik (IHE)
der Universität Karlsruhe (TH) (ISSN 0942-2935)

Fortführung als
"Karlsruher Forschungsberichte aus dem Institut für Hochfrequenztechnik und Elektronik" im Universitätsverlag Karlsruhe
(ISSN 1868-4696)

Strategies for Time Domain Characterization of UWB Components and Systems

Zur Erlangung des akademischen Grades eines

DOKTOR-INGENIEURS

von der Fakultät für
Elektrotechnik und Informationstechnik
der Universität Fridericiana Karlsruhe (TH)

genehmigte

DISSERTATION

von

Dipl.-Ing. Elena Pancera
aus Verona, Italien

Tag der mündlichen Prüfung: 15.05.2009
Hauptreferent: Prof. Dr.-Ing. Dr. h.c. Dr.-Ing. E.h. Werner Wiesbeck
Korreferent: Prof. Stefano Maci

Preface

This Dissertation has been developed and written during my time as research assistant (wissenschaftliche Mitarbeiterin) at the Institut für Hochfrequenztechnik und Elektronik (former Institut für Höchstfrequenztechnik und Elektronik) at the Universität Karlsruhe (TH).

Firstly I want to give my thanks to Prof. Dr.-Ing. Dr.h.c. Dr.-Ing. E.h. Werner Wiesbeck for his outstanding support to this work, for the precious scientific discussions and for having been the Hauptreferent of my Dissertation.

I want to give my thanks to Prof. Dr.-Ing. Thomas Zwick, Director of the IHE, for his rigorous and accurate support and his helpfulness for scientific discussions.

I also want to give my thanks to Prof. Stefano Maci, Università degli Studi di Siena, Italy, for having been the Korreferent of my Dissertation.

Moreover, I want to thank all my colleagues and in particular my room colleagues for the stimulating work-atmosphere. A particular thanks to the UWB-Team of the IHE for the inspiring and motivating scientific considerations. In addition I want to thank Dipl.-Ing. Christian Sturm for the critical re-reading of this Dissertation. Besides, I want to thank the students that with their Diplomarbeit and Masterarbeit have provided some elements to perform this work.

Finally, I thank my family for having supported me during all the time.

Elena Pancera

Karlsruhe, May 2009.

Contents

Index of Used Acronyms and Symbols

In this section, the most important acronyms and symbols used in this thesis are reported. It is highlighted that in the thesis the same character is used for real and complex quantities. Moreover, the character a indicates a scalar quantity (real of complex) while **a** indicates a vectorial quantity. The time domain quantities are written with small letter while the frequency domain quantities are written with capital letters.

Acronyms

AUT	Antenna Under Test
BP	Bandpass
CPW	Coplanar Waveguide
DFT	Discrete Fourier Transform
DUT	Device Under Test
EIRP	Effective Isotropic Radiated Power
EU	European Union
FCC	Federal Communications Commission
FD	Frequency Domain
FT	Fourier Transform
FWHM	Full Width at Half Maximum
GDT	Group Delay Time
IDFT	Inverse Discrete Fourier Transform
IEE	The Institution of Electrical Engineers
IEEE	Institute of Electrical and Electronic Engineers
IFT	Inverse Fourier Transform
ISM	Industrial, Scientific and Medical radio bands
JRC	Joint Research Center, Ispra, Italy
LOS	Line of Sight
LTCC	Low Temperature Co-fired Ceramic
MF	Matched Filter
NB	Narrow-Band
NLOS	Non Line of Sight
PSD	Power Spectral Density
RCS	Radar Cross Section
SNR	Signal to Noise Ratio
SVD	Singular Value Decomposition
TD	Time Domain
TEM	Transversal Electromagnetic Wave
UWB	Ultra Wideband
VNA	Vector Network Analyzer
VSWR	Voltage Standing Wave Ratio
WLAN	Wireless Local Area Network

Latin Symbols

B	Signal Bandwidth
d	Distortion
$\mathbf{e}$	Electric field vector in the time domain
$\mathbf{E}$	Electric field vector in the frequency domain
f	Frequency
f_c	Centre frequency
f_{cutoff}	Cut-off frequency
f_H	Higher cut-off frequency
f_L	Lower cut-off frequency
F	Fidelity
G	Gain of an antenna
h	Impulse response (scalar), i.e. in a particular direction
$\mathbf{h}$	Impulse response vector
$[\mathbf{h}]$	Polarimetric impulse response matrix
H	Transfer function (scalar), i.e. in a particular direction
$\mathbf{H}$	Transfer function vector
$[\mathbf{H}]$	Polarimetric transfer function matrix
j	Imaginary unit $j^2 = -1$
N	Noise power
P	Peak
r	Distance
R	Cross-correlation
$[\mathbf{S}]$	Scattering matrix
S_{11}	Input reflection parameter
S_{12}	Reverse transmission parameter
S_{21}	Forward transmission parameter
S_{22}	Output reflection parameter
t	Time
T	Signal time duration
u	Voltage signal in the time domain
U	Voltage signal in the frequency domain
Z	Impedance

Greek Symbols

α	Percent of ringing duration
γ_τ	Mixed Gaussian shape
δ	Dirac Impulse
ε_r	Dielectric constant of a substrate
η	Polarization component
θ	Elevation angle
ϑ	Offset of a Dihedral from the vertical
λ	Wavelength
λ_g	Guided wavelength
μ	Mean value
ξ	Realization of a Gaussian random process
ζ	Polarization component
Ξ	Gaussian random process
ρ	Ripple
σ	Standard deviation
$[\sigma(f)]$	Radar Cross Section matrix in the frequency domain
$[\sigma(t)]$	Radar Cross Section matrix in the time domain
τ	Time difference
τ_g	Group Delay Time
τ_{FWHM}	Full Width at Half Maximum
$\tau_{r,\alpha}$	Ringing duration at the percent α of the peak
ψ	Azimuth angle
ω	Angular frequency

Operators and Mathematical Symbols

r	Scalar quantity
$\mathbf{r}$	Column vector
$\mathbf{r}^{\mathrm{T}}$	Row vector (transpose of $\mathbf{r}$)
$\hat{\mathbf{r}}$	unit vector parallel to the vector $\mathbf{r}$
$\hat{\mathbf{r}}_\theta$	Local coordinate system in θ-direction
$\hat{\mathbf{r}}_\psi$	Local coordinate system in ψ-direction
$|\mathbf{r}|$	Absolute value of $\mathbf{r}$
$\angle \mathbf{r}$	Phase of $\mathbf{r}$
$< \mathbf{r}_1, \mathbf{r}_2 >$	Scalar product between $\mathbf{r}_1$ and $\mathbf{r}_2$
$[\mathbf{r}]$	Matrix
$\Re[\cdot]$	Real part operator
$*$	Convolution operator
$*^{-1}$	Deconvolution operator
H^+	Analytic signal of H
H^*	Complex conjugate of H
$\|\mathbf{H}\|_p$	p-norm of $\mathbf{H}$
$\dot{u}$	time derivative of u
$\mathcal{H}[\cdot]$	Hilbert transform
$\mathcal{F}[\cdot]$	Fourier transform
$\mathcal{F}^{-1}[\cdot]$	Inverse Fourier transform
$\exp$	Exponential function
$\max$	Maximum
$\min$	Minimum
∞	Infinity

General Symbols

A	Antenna
co	Copolarization
cross	Crosspolarization
h	Horizontal polarization
rand	Random
Rx	Receiver
Tx	Transmitter
v	Vertical polarization

Constants

c_0 Velocity of light in vacuum: $2.9979 \cdot 10^8$ m/s
e number of Euler: 2.718
ε_0 Permittivity of the vacuum: $8.854 \cdot 10^{12}$ As/(Vm)
μ_0 Permeability of the vacuum: $4\pi \cdot 10^{-7}$ Vs/(Am)
π Pi: 3.1415...

1 Introduction

1.1 UWB Technology

Ultra Wideband (UWB) is a wireless trasmission technology which is characterized by the usage of a very large bandwidth. The occupation of a huge frequency band makes the UWB technology a powerful solution for many applications and brings many advantages with respect to classical narrowband (NB) systems. In the multimedia case, for example, UWB permits the transmission of extremely high data rates. The huge bandwidth also results in high resolution for Radar applications, such as for localization purposes or medical purposes, among them: investigations of internal organs with high resolution, cancer recognition, etc. Moreover, also sensor networks can take advantages from this technology, since the huge bandwidth allows for the coexistence of a large number of users. Furthermore, the usage of a large bandwidth permits to achieve high information capacity, according to Shannon's [1] relationship of channel capacity, bandwidth and signal to noise ratio, being the channel capacity directly proportional to the bandwidth.

The first attestation of Ultra Wideband transmission has been given by Heinrich Hertz in 1887 in Karlsruhe, when he was able to transmit for the first time short pulses over a distance of several meters. After these first experiments narrowband transmission has achieved more interest and only at the end of the 1950s the UWB technology has been investigated for communications, Radar and related applications, mainly for special purposes and military applications. However, a renewed interest in UWB for commercial and consumer purposes started to appear in the 1980s. Due to its manifold advantages and benefits compared to classical narrowband technology, including high data rate, low transmit power and low interference level, in recent times the UWB technology has received increasing interest in the research field and many components and systems for UWB applications have been developed and presented in literature.

A first regulation of a UWB frequency band for commercial and consumer applications has been performed in 2002 by the Federal Communications Commission (FCC) in the USA [2], which has regulated the usage of UWB devices allocating a frequency band of 7.5 GHz between 3.1 and 10.6 GHz for communications purposes with a very low power spectral density of -41.3 dBm/MHz. Recently (February 2007) also the European Union (EU) has approved a UWB frequency spectrum in Europe [3] allocating a band of 2.5 GHz between 6 and 8.5 GHz for UWB operations, additionally also the usage of the frequency band between 3.4 and 4.8 GHz is allowed with specific duty cycle restrictions. These different frequency masks are illustrated in Fig. 1.1 for comparison.

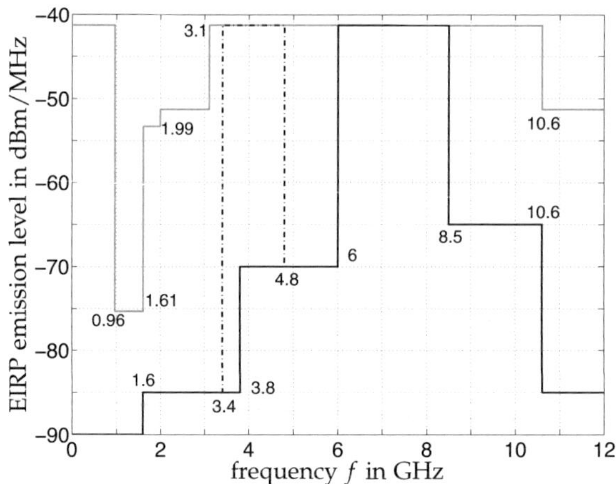

Figure 1.1: The UWB mask: FCC regulation (gray) and EU regulation without (black, solid line) and with duty cycle limitation (black, dotted line).

1.2 State of the Art

The interest in the usage of Ultra Wideband systems for commercial and consumer purposes has only grown in recent years, when there has been more and more demand for high data rate wireless transmission, requirement of more and more capacity, increment of the number of users. This interest in UWB has grown both in academia and industry leading to the increment of publications of new UWB components and systems in literature.

Up to now, many antenna typologies have been proposed, both for communication and Radar applications, including arrays and dual polarized antennas. Strategies have been developed permitting to obtain high gain or to decrease the cross-polarization and coupling in arrays over a large frequency interval [4], [5]. Together with methods for improving the behavior of the antennas, also methods for quantify their performance in a large bandwidth have been proposed, leading to a characterization of UWB antennas in the time domain [6], [7], [8]. Description parameters have been introduced to quantify the antenna distortion [9], [10], [11], group delay time variation [12], and spatial pulse preserving capability [13].

As for antennas, also many analog UWB filter typologies have been proposed in literature. The typologies of UWB filters are either derived from narrowband filter design (such as resonator stubs filters [14], [15]) which have been adapted for the UWB case. In literature many filter prototypes realized with different fabrication techniques (microstrip [16]-[20], CPW [21]-[25], LTCC...) are reported. The behavior of the various developed UWB filter structures is usually only analyzed in the frequency domain, i.e. only the matching of the realized filter to the given UWB spectral mask and its group delay time behavior are taken into considerations. In [26] an analysis of a filter time domain behavior has been performed. In [27], [28], [29] an analysis of the poles and zeros of a filter has been conducted in order to have the possibility to design an approximation of the desired time domain filter mask. In [30] a first UWB analysis of the impact of the filter frequency domain non-idealities on its time domain behavior and pulse preserving capability has been proposed.

UWB filters are mainly used for spectrum restriction according to the regulations and not for spectrum optimization. Conversely, in [31] a UWB based method for spectrum optimization using a particular analog filter structure has been presented. Together with analog filters, investigations have been performed also on digital filters, in particular for creating *ad hoc* UWB signals matching the required UWB mask [32]-[34].

In literature, also examples of an integration of UWB antennas and filters are reported and different methods for performing the integration are addressed [35], [36]. In [37], [38] a detailed analysis of these different methods is reported.

Investigations of UWB systems have been made not only for communications purposes, but also for ranging and localization applications conducting to an increment of the research activities in the UWB Radar technology [39], [40], [41]. Contributions on UWB Radar, including both hardware (antennas, arrays, …) and software (imaging algorithms, …) can be found in literature. Measurements of Radar cross sections both in the frequency domain and in the time domain are reported [42]. Also calibration methodologies have been presented. In [43] a first example of a dedicated UWB calibration method in the time domain has been reported.

1.3 Goal of the Thesis

The extremely high bandwidth that is available from the new UWB regulations sets completely new challenges regarding component and system design. Moreover, in order to describe and analyze the behavior of the involved components, new description methodologies and performance criteria have to be developed.

In the usually applied methods in RF engineering, devices (e.g. filters, antennas...) are described and characterized through an analysis conducted in the frequency domain. The aim of this thesis is to perform a complete analysis of the UWB system components in the time domain, both regarded as single components and from a system point of view. This permits to directly assess the dispersion behavior of the involved devices. Moreover, this time domain analysis is supported by a correlation analysis in the time domain, which permits to quantify the distortion introduced by the device on the signal. By conducting these two analyses together, it is possible to have a clear quantification of the deviation of the behavior of the device from the ideally expected behavior, making it possible to introduce correction strategies. Furthermore, with the performed investigations, it is possible to quantify the impact of the frequency domain non-idealities on the time domain behavior.

Moreover, in order to verify the proposed criteria, prototypes of hardware components (filters, integration of antenna and filter) have to be developed and fabricated.

From measurements of the constructed prototypes the proposed criteria are validated.

Together with the analysis of the single components, the aim is to also perform a complete analysis from a system point of view, both of the UWB radio link and of the UWB Radar link. Performance criteria have to be developed to optimize the behavior of the UWB system and correction strategies have to be introduced from a system point of view, for both the UWB radio link and the UWB Radar link. Also in this case, the developed criteria and correction strategies have to be validated through measurements of particular UWB scenarios.

1.4 Novelty of the Thesis

The results of this thesis contribute to improve the performance of UWB components and systems in the time domain. This is achieved by the introduction of criteria for quantifying the time domain performance of both UWB components and systems. This approach permits to have a clearer insight into the distortion and the dispersiveness in the time domain caused by the non-ideal behavior of the UWB devices and UWB systems. Using the developed criteria also correction strategies are introduced, which allow to improve and optimize the behavior of UWB systems.

In the analysis of UWB filters, (ref. to chapter 3), using a statistical approach, a novel method to quantify the impact of the filter frequency domain non-idealities (non-flat filter frequency mask and non-constant filter group delay time) on the filter time domain behavior has been derived. With this method bounds on the standard deviations of the passband filter mask and of the filter group delay time can be found, which, if respected, permit to have only negligible distortion of the transmitted signal with respect to the case of an ideal filter.

UWB filters are also analyzed when integrated with UWB antennas (ref. to chapter 4). In this context a particular approach to integrate the filter in the antenna feeding line/ground plane is taken into consideration and compared to the filter integration in the antenna radiating element. From the conducted analysis it is concluded that the integration in the antenna ground plane/feeding line has better performance with respect to the integration in the antenna radiating element.

In a UWB radio system link analysis (ref. to chapter 5), regarding the signal at the receiver, novel criteria are found that permit to assess the quality of the UWB link from the perspective of the receiver including the non-ideal system behavior. This approach leads to the definition of two quality criteria which permit to quantify the worsening of the signal to noise ratio of a UWB radio link due to the non-ideal behavior of the antennas.

By investigating correction strategies for the optimization of UWB system performance, in terms of matching and fulfillment of the required mask (ref. to chapter 6), a new design approach for analog filters is developed. This filter design permits to realize special structures that allow both to select the required frequency interval and to pre-distort the signal in order to compensate for the spectral non-idealities of the transmitter components. With this filter design the transmit power can be maximized and hence an improvement of the signal to noise ratio at the receiver can be achieved.

Finally, through an investigation of the UWB Radar link in the time domain, a novel calibration procedure for UWB Radar is introduced (ref. to chapter 7), which is entirely developed in the time domain and requires very low computational effort and only 4 measurements per calibration target and does not require an "empty room" calibration.

1.5 Organization of the Thesis

In order to achieve the stated goals, the thesis has been organized in two different stages. In the first stage the single UWB components (filters, antennas integrated with filters) are separately investigated, in order to assess their time domain behavior and dispersiveness. In the second stage the single elements, previously separately investigated, are analyzed from a system point of view, performing an investigation of the UWB radio link and of the UWB Radar link.

First of all, the mathematical basis for a frequency domain analysis, a time domain analysis and a correlation analysis is given in chapter 2. Here, firstly the commonly used frequency domain analysis is regarded and the most important parameters are introduced. Secondly, the time domain analysis of UWB components is discussed and performance parameters, which permit to quantify the time domain behavior of UWB components, are presented. Finally, specific parameters are introduced with the correlation analysis, which allow to assess the pulse preserving capability and the distortion caused by a UWB component.

These analyses are applied to UWB filters in chapter 3. Here, a complete investigation of the impact of the non-ideal filter behavior in the frequency domain (in terms of both non-flat passband filter frequency mask and non-constant filter group delay time) on the filter time domain behavior is performed. In doing that, performance criteria are introduced, which can be easily measured and are helpful for practical UWB filter design. Moreover, various approaches to the design and hardware implementation of UWB filters are regarded. The particularities and characteristics of the different realizations are discussed and prototypes are fabricated. From measurement results these prototypes are characterized in the frequency domain and in the time domain. Furthermore, the developed theory for the impact of the frequency domain non-idealities on the time domain behavior of UWB filters is validated.

After the analysis of UWB filters as single element, an investigation of the integration of filters and antennas for UWB applications is performed in chapter 4. In doing that, firstly, the performance parameters of UWB antennas are introduced and different approaches to the integration of filters and antennas are discussed and compared. Prototypes for different realization techniques are developed and fabricated. Then, the behavior of the fabricated antenna+filter devices is investigated both in the frequency domain and in the time domain.

Once the single UWB components have been analyzed separately, they are investigated from a system point of view. In chapter 5 the analytical description of a complete UWB radio link is regarded, both in the frequency domain and in the time domain.

Here, from the receiver point of view, the condition for obtaining the optimum signal to noise ratio is investigated. This leads to a fidelity criterion in the time domain of the distortion that the transmitted signal undergoes in the different radiation directions with respect to the signal transmitted in the main beam direction.

In chapter 6 strategies for optimizing the transmitted signal power are presented. These strategies consist in compensating for the non-ideal spectral behavior of the UWB components. A mathematical approach for the optimization of the radiated signal spectrum is presented. This optimization is conducted in terms of optimum match of the spectrum of the radiated wave to the required UWB mask. Dedicated filter structures that allow to shape the spectrum of the radiated signal are developed, fabricated and measured.

Similarly to chapter 5, where the UWB radio link is discussed, in chapter 7 the UWB Radar link is regarded. Firstly, a fully polarimetric analytical time domain description of the UWB Radar link is presented. Secondly, a fully polarimetric UWB Radar calibration entirely performed in the time domain is shown and validated through measurement results.

Throughout the thesis, all theoretically derived relationships between frequency domain and time domain and all the developed performance criteria are validated through measurements on fabricated hardware prototypes.

In the introduction of each chapter the handled problem and the goal of the chapter are discussed and also the state of the art of the solutions to the given problem are illustrated, while the conclusion contains the most important results of the chapter and the novelty introduced in the chapter with respect to the state of the art. At the end of the thesis the obtained results to the stated problems and the achievement of the proposed goals are discussed in the conclusions. An Appendix is also inserted, to summarize specific mathematical background used in the thesis.

2 Theoretical Background

As introduced in the previous chapter, the main focus of this thesis is analyzing UWB system components regarding their behavior in the time domain and the effect of the component's non-idealities. This concerns in particular the effect on the pulse preserving capability of the components. First of all the mathematical basis of the analyses that are presented in the following chapters will be introduced. In literature, components (filters and antennas) are usually characterized by an analysis conducted mainly in the frequency domain. In this thesis, a time domain analysis for UWB systems will be developed, which gives a clearer insight into distortion of the time domain signals. Moreover, time domain parameters for performance assessment will be defined, that are directly related to typical signal distortion. Also interrelations between frequency domain analysis and time domain analysis and parameters will be regarded.

In the following, firstly the commonly used frequency domain analysis is discussed and the most important parameters are introduced. Then, the time domain analysis for UWB components is regarded. Also in this case, performance parameters are presented, which permit to characterize the device's behavior in the time domain. Moreover, in order to quantify the pulse preserving capability of the device itself, i.e. the distortion that the device introduces on the signal, specific parameters are introduced.

2.1 Frequency domain analysis

In this section, the main parameters for the frequency domain analysis of UWB devices are presented.

The frequency domain behavior of a two-port device (see Fig. 2.1) is usually described by its transfer function $H(f)$ [44]. Letting $U_{\text{in}}(f)$ be the complex voltage signal, $U = |U|e^{j\angle U}$ with $\angle U$ the phase of U, at the device's input and $U_{\text{out}}(f)$ the complex voltage signal at the device's output in the frequency domain, the input-output relationship of the two-port can be written as

$$H(f) = \frac{U_{\text{out}}(f)}{U_{\text{in}}(f)} \tag{2.1}$$

with $H(f)$ being the transfer function of the device. In case of a typical two-port device (e.g. filters), the transfer function corresponds to the S_{21} parameter of the two-port device. It has to be pointed out that for particular devices (e.g. antennas) it is not only a function of the frequency, but it depends also on the angular directions, as it is illustrated in chapter 5.

In the following sub-section the most important frequency domain parameters, which can be derived directly from $H(f)$ are described.

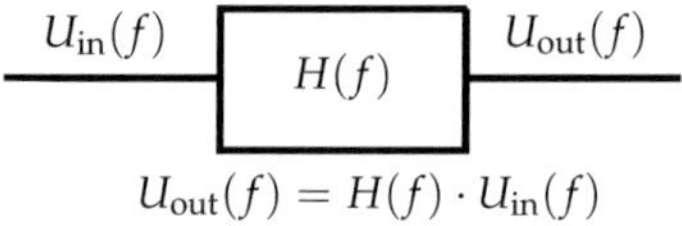

$$U_{\text{out}}(f) = H(f) \cdot U_{\text{in}}(f)$$

Figure 2.1: Two-port device: frequency domain description.

2.1.1 Frequency domain parameters

The most important parameters that permit to characterize and evaluate the performance of the device in the frequency domain can be directly calculated from the device's transfer function $H(f)$.

To derive these parameters, firstly the previous equation is rewritten in complex notation

$$H(f) = |H(f)| \cdot e^{j\angle H(f)} \tag{2.2}$$

where $|H(f)|$ represents the amplitude and $\angle H(f)$ the phase of the device's transfer function. From these two terms, the transfer function amplitude and its phase, it is possible to define the most important frequency domain parameters.

Amplitude $|H(f)|$ This parameter permits to describe the impact of the device on the amplitude of the incoming signal, i.e. how the different frequency components of the incoming signal are weighted by the device. For the analysis of many kind of devices (such as filters), the knowledge of this weighting permits to assess the frequency mask of the device itself. In other cases (e.g. antennas), the commonly analyzed parameter is the squared transfer function absolute value, from which it is possible to derive other related parameters such as the gain.

Group Delay Time $\tau_g(f)$ Starting from the phase term, it is possible to define another important frequency domain parameter: the Group Delay Time (GDT). It is defined as

$$\tau_g(f) = -\frac{1}{2\pi} \frac{\partial \angle H(f)}{\partial f} \, . \tag{2.3}$$

The GDT gives an estimation of the distortion of the signal, introduced by phase differences for different frequencies. A non constancy of the group delay indicates a resonant character of the device. This means that the structure is able to store energy.

Both Amplitude $|H(f)|$ and Group Delay Time $\tau_g(f)$, are not single value parameters but functions over frequency, which makes it impossible to quantify the resulting distortion in one single number. In the following it will turn out that, when the analysis is performed in the time domain, this will result in simple parameters that allow for performance evaluation with one single number. Moreover, different distortion effects can be clearly identified and distinguished.

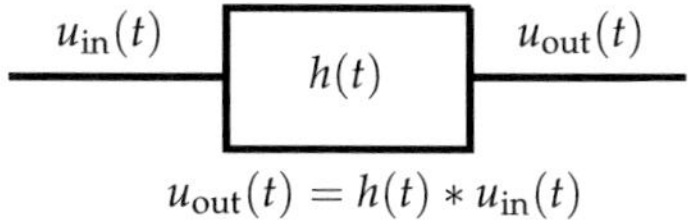

$$u_{\text{out}}(t) = h(t) * u_{\text{in}}(t)$$

Figure 2.2: Two-port device: time domain description.

2.2 Time domain analysis

In this section, the main parameters for the time domain analysis of UWB devices are presented. The starting point of the analysis in the time domain is the device's impulse response $h(t)$. It can be calculated from the frequency domain analytic transfer function[1], defined as [45]

$$H^+(f) = \begin{cases} 2H(f) & f > 0 \\ H(f) & f = 0 \\ 0 & f < 0 \end{cases} \tag{2.4}$$

where the factor 2 is used in order to conserve the energy of the signal, and then taking the real part ($\Re[\cdot]$) of its inverse Fourier Transform (IFT, $\mathcal{F}^{-1}$), namely

$$h(t) = \Re\left[\,\mathcal{F}^{-1}[H^+(f)]\,\right] . \tag{2.5}$$

The absolute value $|h^+(t)| = |\mathcal{F}^{-1}[H^+(f)]|$ gives the envelope of the impulse response.

The time domain relationship between the input signal $u_{\text{in}}(t)$ and the output signal $u_{\text{out}}(t)$ (see Fig. 2.2) is defined through the convolution operation, namely

$$u_{\text{out}}(t) = h(t) * u_{\text{in}}(t) \tag{2.6}$$

Hence, it can be recognized that the device's impulse response is directly obtained in the time domain as output of the device when a very short pulse (in the ideal case a Dirac-impulse) is applied at the input.

2.2.1 Time domain parameters

In the time domain there are three main parameters that permit to evaluate the device's behavior in one single number. They are related to the device's impulse response and its envelope [7], [46].

Peak Amplitude P It represents the peak value of the device's impulse response envelope, namely

$$P = \max_t |h^+(t)| . \tag{2.7}$$

It is a measure for the maximal value of the strongest peak of the time domain impulse response envelope of the device and hence a direct measure for the peak

---

[1]It has to be noted that a direct application of the Inverse Fourier Transform to the device's transfer function gives a complex-valued impulse response, which is not of physical meaning for an excitation of the device with real valued signals.

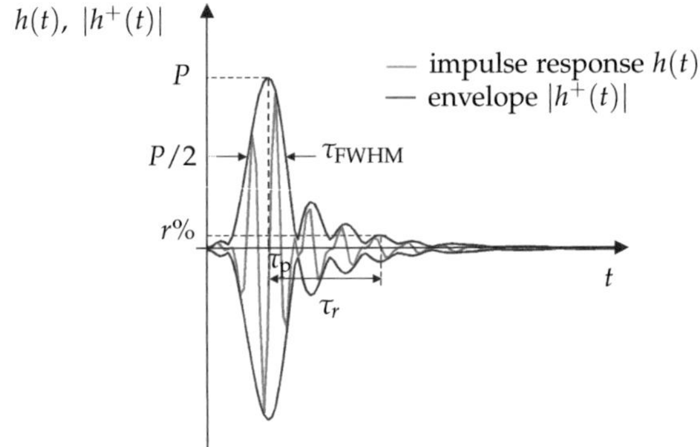

Figure 2.3: Filter impulse response, absolute value of its complex envelope and time domain parameters.

power that can be achieved for impulse radiation. A high peak value P is desirable.

Full Width at Half Maximum τ_{FWHM} It is mathematically defined as

$$\tau_{\mathrm{FWHM}} = t_1\big|_{|h^+(t_1)|=P/2} - t_2\big|_{t_2<t_1,\, |h^+(t_2)|=P/2} \,. \tag{2.8}$$

It describes the broadening of the transmitted pulse due to the non-ideal behavior of the device and it is thus a direct measure for the dispersiveness of the device.

Ringing τ_r Of interest is the ringing duration τ_r, which is defined as the time until the envelope has fallen from the peak value P below a fraction r of the peak amplitude, namely

$$\tau_r = t_r\big|_{|h^+(t_r)|=rP} - t_P\big|_{t_P<t_r,\, |h^+(t_P)|=P} \,. \tag{2.9}$$

Ringing is typically caused by temporary energy storage in resonant structures. Hence, the ringing duration is a direct measure for the energy storage capability of the device. The ringing effect is undesired. The energy contained in ringing is not useful for communications, Radar or sensing and lowers the peak value P.

Hence, in contrast to the frequency domain analysis, in the time domain analysis the performance of the device can be assessed with three parameters that result in one single number and have a direct relation to important properties of the device e.g. pulse power preservation and dispersiveness.

These three parameters are illustrated in Fig. 2.3 for a generic impulse response.

In Tab. 2.1 the parameters of the time domain and frequency domain analyses are summarized.

Table 2.1: Frequency Domain-Time Domain Description Parameters.

	Frequency Domain	Time Domain
Function	Transfer Function $H(f)$	Impulse Response $h(t)$
Related Function	Analytic Transfer Function $H^+(f)$	Envelope $|h(t)^+|$
Parameter	$|H(f)|$, $\tau_g(f)$	P, τ_{FWHM}, τ_r

2.3 Fidelity Analysis

In addition to the parameter analysis in the frequency domain and in the time domain, also a correlation analysis in the time domain is regarded, in order to evaluate the amount of distortion introduced by the device on the signal. Also the result of the correlation analysis can be described in one single number. From the knowledge of the amount of this distortion, it is possible to improve the system itself, as it will be shown in chapter 5. In the following, the mathematical basis for the correlation analysis in the time domain is given.

In the performed analysis, it is of interest to quantify the variation of the device's output signal with respect to a reference signal (for example the ideally expected output). A parameter that permits to describe the variation between two signals $u_1(t)$ and $u_2(t)$ is the so-called distortion d [47]

$$d = \min_{\tau} \int_{-\infty}^{+\infty} \left| \frac{u_2(t+\tau)}{||u_2(t)||_2} - \frac{u_1(t)}{||u_1(t)||_2} \right|^2 dt \tag{2.10}$$

where $|| \cdot ||_2$ represents the 2-norm[2]. The two signals have been normalized by the 2-norm in order to have unit energy and the delay τ has been varied in order to find the minimum of the integral. If the previous integral is expanded, omitting the terms that do not influence its minimum, it reduces to

$$d = \min_{\tau} \left\{ 2 \cdot \left[1 - \underbrace{\int_{-\infty}^{+\infty} \frac{u_2(t+\tau)}{||u_2(t)||_2} \cdot \frac{u_1(t)}{||u_1(t)||_2} dt}_{F} \right] \right\} = \min_{\tau} \left\{ 2 \cdot [1 - F] \right\} \tag{2.12}$$

where F is usually referred to as fidelity and expresses the cross-correlation between the two signals. Consequently, from the previous equation, it can be inferred that the minimum distortion is obtained when the cross-correlation between the two signals has its maximum value.

---

[2] The 2-norm of a signal $x(t)$ is defined as

$$||x||_2 = \sqrt{\int_{-\infty}^{+\infty} |x(t)|^2 dt} \, . \tag{2.11}$$

It is also important to point out that in the correlation analysis only the shapes of the signals are investigated and no information about the signal energy content is obtained. Consequently, for a comprehensive analysis, together with F also the peak amplitude P has always to be regarded.

In the next chapters this analysis is applied in different scenarios (filters, antennas, Radar) always with the aim to quantify the distortion.

3 Time Domain Analysis of UWB Band Pass Filters

The aim of this chapter is to investigate the time domain behavior of UWB bandpass (BP) filters, that are applied in order to restrict the transmit signal spectrum to the frequency range permitted by the regulations. In particular the goal is to analyze the impact of the filter's non-idealities in the frequency domain on its time domain behavior. The aim is to quantify the distortion effect operated by the filter on the transmitted pulse, i.e., to evaluate the pulse preserving capability of the filter when its behavior in the frequency domain is not ideal. Moreover, it is of interest to find bounds on the non-idealities in the frequency domain that will cause only negligible distortion in the time domain.

In literature there are several investigations on the filter time domain behavior. For example, in [27] and [28] an analysis of the poles and zeros positions of the filter has been conducted in order to have the possibility to design an approximation of a desired time domain filter mask. However, in these contributions a direct relationship between the non-idealities in the frequency domain and the time domain (filter impulse response) has not been derived. In [48], an exhaustive analysis of the approximation of the filter impulse response to a particular function with a pole-zero analysis of the filter transfer function has been published. Moreover, in [29] an example of a filter design in the time domain has been provided using the commensurate network theory, but without investigating the filter non-idealities. An exhaustive analysis of the filter time domain behavior has been performed in [26]. Here the time domain responses for different filter topologies have been examined. However, also in this case, even if a relationship between frequency and time domain has been analyzed, an investigation of the practical filter non-idealities has not been performed.

The aim of this chapter is to propose a new performance analysis that allows to quantify the effect of a non ideal frequency domain behavior on the time domain behavior. The regarded criteria can be easily measured and are helpful for practical UWB filter design applications. The developed analysis has general validity and can be applied to different UWB scenarios (e.g. European UWB regulation, FCC ...). It directly allows for deriving bounds in the frequency domain that, if respected, will guarantee negligible distortion of the time domain filter output signal. Starting from the general formulas, derived practical results are focused on filters for the European UWB regulation.

Moreover, practical realizations of Ultra Wideband filters are investigated. For that purpose different realization techniques are compared and prototypes are fabricated. To these prototypes the novel criteria are applied in order to prove their validity and suitability for practical performance assessments.

The chapter is organized as follows. Firstly, the ideal filter is regarded. Secondly, the design rules for UWB filters are shortly presented, in order to explain the peculiarities of the different realizable filter typologies. Then, the frequency domain non-idealities are described and investigated. Later, their impact on the filter time domain behavior is evaluated proposing a particular statistical analysis. From this analysis, detailed results for the impact of the different non-idealities in the frequency domain on the distortion of the filter output signal are obtained and discussed. Finally, the obtained results are validated through the investigation of the time domain behavior and the pulse preserving capability of different practical UWB filter typologies.

3.1 The Ideal Filter

In the ideal case a bandpass filter presents a rectangular frequency mask, which is flat in the passband and shows a constant group delay time, as illustrated in Fig. 3.1. The flatness of the mask in the passband means that the different frequency components of the incoming signal are weighted all equally, i.e. their amplitude is preserved. Moreover, a constant GDT means that the filter's phase is linear and consequently all frequency components are delayed by the same amount given by the constant value of the GDT. Letting B be the filter's bandwidth, f_c the centre frequency and τ_p its constant GDT, from eq. (2.2) the ideal filter's transfer function can be written as

$$H_{\mathrm{id}}(f) = \mathrm{rect}\left(\frac{f - f_c}{B}\right) \cdot \exp\left[-j2\pi f \tau_p\right] \tag{3.1}$$

where the term $-j2\pi f \tau_p$ represents the phase of the filter due to a constant group delay time of entity τ_p. The filter impulse response can be calculated recalling the IFT relationship

$$\mathrm{rect}\left(\frac{f - f_c}{B}\right) \cdot \exp\left[-j2\pi f \tau_p\right] \quad \circ\!\!-\!\!\!-\!\!\bullet \quad \underbrace{B \, \mathrm{sinc}\left(B \cdot (t - \tau_p)\right) \cdot \exp[j2\pi f_c t]}_{h^+(t)} . \tag{3.2}$$

The envelope $h^+(t)$ of the impulse response is given by the absolute value of a sinc function. Its peak is shifted from the origin by an amount given by the constant value of τ_p, the τ_{FWHM} is determined by the spectral extension of the filter rectangular mask B.

3.1.1 UWB Filter Design

In the following, the basic concepts of the design of filter structures, which are typically used in the realization of filter typologies, are presented, in order to show that realizable filters intrinsically differ from the regarded ideal one.

In filter design, the usually defined goal is the amplitude squared transfer function of the filter, i.e. the goal is usually specified in the frequency domain. The design goal

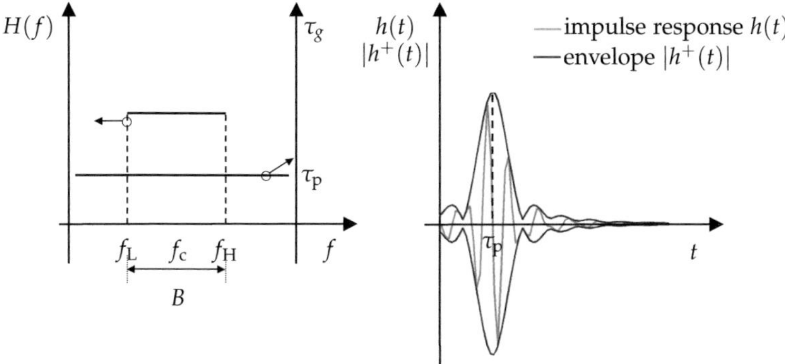

Figure 3.1: Ideal filter frequency mask and group delay (left) and corresponding filter impulse response and its envelope (right).

can be mathematically described as

$$|H(f)|^2 = \frac{1}{1 + \epsilon^2 F_n^2(f)} \tag{3.3}$$

where ϵ is a constant related to the filter ripple, $F_n(f)$ represents the so-called filter characteristic function and n the filter order. According to the selected $F_n(f)$, different filter transfer functions (Butterworth, Chebychev, ...) and hence different filter masks and group delay times can be obtained [14]. Each $F_n(f)$ has particular properties (such as fast cutoff in the transitional band, linear phase, ...) that make the filter suitable for different operations. It can be directly observed that different realization typologies have an intrinsic degree of non-ideality due to the fact that in the filter design process a particular mask is realized, which, according to the selected filter characteristic function and order, presents passband ripple and non-constant GDT. E.g., Buttherworth filters have a constant passband mask but their transitional bands are quite large. On the other hand, Chebychev filters have a ripple in the passband mask and in the GDT. Linear phase filters have constant GDT [14], [15].

In the design of practical filters, there are various degrees of freedom, which permit to construct the filter that satisfies the given requirements at best. The design of filters starts from the identification of the requirements, which usually are:

Frequency Interval i.e. the filter passband;

Filter Characteristic Function it permits to obtain a particular shape of the filter transfer function and a particular GDT;

Filter Order it determines attenuation, sharpness of transitional band (bands), ...

Realization technology lumped elements, distributed elements, microstrip, CPW, LTCC, etc. Each realization has particular impact on the filter performance, e.g. on the efficiency.

Consequently, the various realizable filter transfer functions have different GDT (different ripple according to the selected order and characteristic function) and passband mask (according to the selected parameter ϵ and to the particular F_n). Moreover, the physical filter realization (selected technology, imperfections in the fabrication process, etc.) also introduces a degree of uncertainty in the filter itself, whose actually measured behavior can differ from the theoretical one. Hence, a useful criterion is required, which permits to directly quantify the filter time domain behavior and its deviation with respect to the ideal case directly from information in the frequency domain, i.e. from the measured data. This criterion has also to be general, i.e. it has to be possible to apply it to different filter typologies, so that it permits, given generic filter measurement data, to directly quantify the filter time domain behavior in term of distortion with respect to the ideal case.

3.2 Filter Non-Idealities in the Frequency Domain

In this section the description of the filter frequency domain non-idealities is presented, i.e. how the filter differs from the ideal case. Moreover the effects of these non-idealities on the filter frequency and time domain behavior with respect to the ideal filter are discussed.

3.2.1 Group Delay Time

The first non-ideality that is taken into consideration is a non-constant GDT. As discussed before, the GDT describes the delay suffered by the different components of the incoming group wave traveling into the filter structure. In the ideal case, the filter presents a linear phase and a constant group delay time, which means that each frequency is delayed by the same time amount. In the realistic case, the GDT is not constant. A non-constant group delay time means that the different components of the incoming group wave are delayed by different Δt and hence the time domain response is distorted. This issue is illustrated in Fig. 3.2. Here, for simplicity and without loss of generality, it is assumed that the input signal is given by the superimposition of two sinusoidal functions with different frequencies, i.e., it has only two frequency components f_1 and f_2. Moreover, in order to better clarify the effect of a non constant GDT, it is assumed that $f_1, f_2 \in B$, with B the filter bandwidth and that the filter mask is ideally flat[1]. In the case of a constant GDT, the two different frequency components of the signal are both delayed by the same amount and consequently in the time domain the output signal is simply a shifted version of the input signal, delayed by the constant amount of the GDT τ_p. On the other hand, if the GDT is not constant, each component

---

[1]This simplification has been assumed to investigate only the effect of a non constant GDT.

suffers a different delay, i.e. the first component is delayed by τ_{p1} and the second one by τ_{p2}. Hence, the output signal is not the delayed version of the input signal anymore, as in the ideal case, but a distorted version, as illustrated Fig. 3.2 (right).

Consequently, the possibility to quantify the distortion suffered by the input signal due to a non constant GDT is an important task when evaluating the performance of filters, in particular of UWB filters. In that case, in contrast to narrow-band filters, due to the huge bandwidth of UWB filters, the constancy of the GDT is a challenging task.

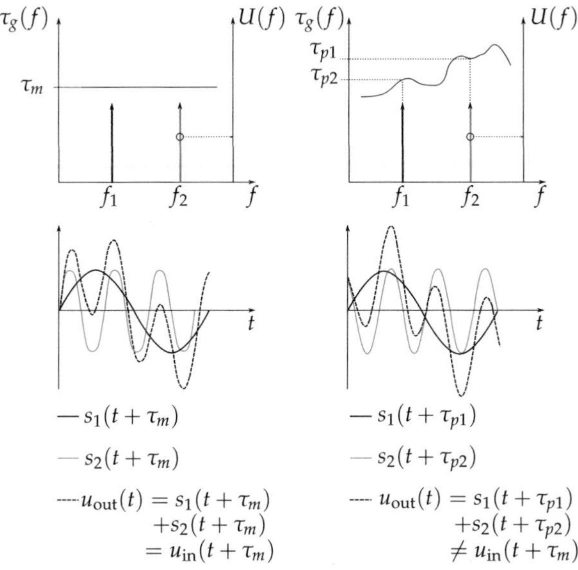

Figure 3.2: Effect of a non-ideal GDT on the filter output signal when the input is the sum of two sinusoidal functions with different frequencies. Left: the ideal case (flat GDT). Right: distortion of the output signal due to a non-constant GDT.

3.2.2 Filter frequency mask amplitude

The second non-ideality in the frequency domain that is investigated is the ripple of the filter mask in the passband. In the ideal case the filter mask has a rectangular shape with constant value in the filter passband frequency range. This constant mask preserves the signal amplitude in the passband frequency interval and consequently the effect of the filter is only to restrict the spectrum, since all frequency components of the input signal in the passband are weighted by the same constant amount and all the other are suppressed. In the non-ideal case the filter passband mask is not constant, i.e. it shows a ripple. This ripple produces a distortion in the filter output signal since the different frequency components of the signal spectrum in the passband are weighted differently. This means that some frequency components are lowered with respect to

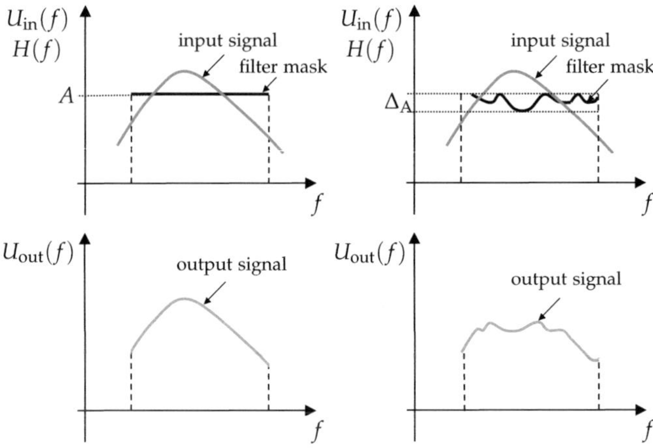

Figure 3.3: Effect of a non-ideal filter frequency mask. Left: the ideal case. Right: the presence of a ripple of entity Δ_A in the passband filter mask.

neighbouring frequency components with less attenuation. The resulting effect is that the amplitude relation versus frequency is deteriorated. This effect is illustrated in Fig. 3.3. Consequently, an investigation of the influence of the ripple in the filter frequency mask on the filter transient behavior is an important topic to be investigated.

3.3 Analysis of the Impact of the Filter Frequency Domain Non-Idealities on its Time Domain Behavior

In order to quantify the impact of the filter non-idealities on the time domain behavior, the filter impulse response is regarded. A method that permits to quantify the deviation of the filter behavior in the time domain, when non-idealities in the frequency domain are present, from the ideal case (i.e. when there is not any non-ideality in the frequency domain) is to evaluate the distortion between the ideal filter impulse response and the measured filter impulse response, resulting from a non-ideal frequency domain behavior. Hence, in order to quantify this distortion, a correlation analysis is performed. Moreover, as stated in the previous chapter, since the correlation analysis alone cannot give any information about the energy content, also the time domain parameters (defined in section 2.2.1) are calculated. Consequently, together with the value of the fidelity F, also the peak amplitude P, the FWHM τ_{FWHM} and the ringing τ_r are calculated.

In order to be able to derive a general relationship between the frequency domain and the time domain behavior, random non-ideal frequency domain transfer functions are stochastically modeled. In this modeling, the non-idealities are described by statistical parameters like e.g. standard deviation.

With the performed analysis, it is possible to find bounds on the values of the statistical non-idealities in the frequency domain that guarantee a quasi-ideal time domain impulse response. Moreover, with the performed analysis it is possible to directly characterize the time domain behavior of a particular filter and its pulse preserving capability from the knowledge of the statistical properties of the non-idealities in the frequency domain from measured data.

3.3.1 Statistical Analysis Procedure

In order to quantify the impact of the deviation of the time domain filter behavior from the ideal case due to the presence of frequency domain non-idealities (non constant GDT and non-flatness of the filter frequency mask), a statistical analysis approach has been developed, since there is no deterministic solutions. In this section, the procedure followed for this statistical analysis is presented.

The effect of both non-idealities is jointly evaluated by creating a high number of random non-ideal filter transfer functions $H_{\mathrm{rand}}(f)$ in the following way. For each random transfer function, firstly, a mask $A_{\mathrm{rand}}(f)$ is generated with a random ripple in the passband. Secondly, a non constant GDT $\tau_g(f)$ is randomly created and then integrated over the frequency (see eq. (2.3)) to obtain the phase term $\angle H_{\mathrm{rand}}(f)$ of the transfer function. Then, according to eq. (2.2) the complete transfer function $H_{\mathrm{rand}}(f)$ of this non-ideal random filter is calculated as

$$H_{\mathrm{rand}}(f) = A_{\mathrm{rand}}(f) \cdot \exp[j\angle H_{\mathrm{rand}}(f)] . \qquad (3.4)$$

From this transfer function, following the procedure described in the section 2.2 in the previous chapter, the corresponding filter impulse response $h_{\mathrm{rand}}(t)$ is derived. In order to quantify the deviation of the obtained impulse response from the ideal case, the values of the peak amplitude P, the FWHM τ_{FWHM} and the ringing τ_r are calculated. Moreover, a correlation analysis has been performed. Being the aim the quantification of the distortion between the ideal filter impulse response $h_{\mathrm{id}}(t)$ and the random one $h_{\mathrm{rand}}(t)$, the fidelity F between $h_{\mathrm{id}}(t)$ and $h_{\mathrm{rand}}(t)$ is evaluated, following the procedure presented in section 2.3 in chapter 2.

3.3.1.1 Random Model for the filter mask

The created filter mask $A_{\mathrm{rand}}(f)$ presents a random ripple, hereafter called $\rho_A(f)$. It is assumed that the values of this ripple are randomly distributed according to a Gaussian distribution and that the ripple follows a Gaussian random process[2] $\Xi_A(f)$. This choice has been made based on [26] and on measurements of real filter implementations. This means that a generic ripple is a particular realization $\xi_i^A(f)$ of the process $\Xi_A(f)$, i.e.

$$\rho_A(f) = \xi_i^A(f) . \qquad (3.5)$$

---

[2]In the Appendix the mathematical background for random Gaussian processes is given.

Such a process is characterized by the mean value μ_A, the standard deviation σ_A and the correlation frequency separation $\Delta f_{\text{corr,A}}$, which gives the rate of variation of the process itself.

Letting the filter passband be in the frequency interval $f_L < f < f_H$, the random filter mask $A_{\text{rand}}(f)$ is generated as[3]

$$A_{\text{rand}}(f) = \begin{cases} 1 - |\rho_A(f)| & f_L \leq f \leq f_H \\ 0 & f < f_L, f > f_H \end{cases} \tag{3.6}$$

$A_{\text{rand}}(f)$ has been set to zero outside the interval $f_L \leq f \leq f_H$ since the interest is to investigate the passband non-idealities.

In the performed analysis the following assumptions are made:

mean value $\mu_A = 0$;

standard deviation σ_A has been taken as the parameter to be varied;

correlation frequency separation $\Delta f_{\text{corr,A}}$. This value has been selected after experimental verifications of different filter implementations[4] (for EU-UWB it has been assumed $\Delta f_{\text{corr,A}} = 0.2$ GHz.)

3.3.1.2 Random Model for the filter GDT

In practical realizations the GDT $\tau_g(f)$ is not constant but presents peaks at the lower and upper frequencies of the passband interval, due to resonance effects of the filter structure at the cutoff frequencies [26]. According to that, the simulated GDT is assumed to have two peaks, one for each transitional band. The shape of each peak has been assumed to be Gaussian. Hence, the part of the random GDT caused by resonance effects results to have a composed Gaussian shape, namely

$$\gamma_\tau(f) = \tau_{m1} \cdot \exp\left[-\frac{(f - f_L)^2}{2\sigma_{G1}^2}\right] + \tau_{m2} \cdot \exp\left[-\frac{(f - f_H)^2}{2\sigma_{G2}^2}\right] \tag{3.7}$$

where τ_{m1}, τ_{m2} are factors that permit to vary the entity of the peaks and σ_{G1}, σ_{G2} are the apertures of the Gaussian curves. After experimental verifications and simulations and from [26] it has be seen that, setting $\sigma_{G1} = \sigma_{G2} = 0.2$ GHz, $\gamma_\tau(f)$ represents typical practical UWB filters for the EU regulation. It has been observed that, for smaller values of $\sigma_{G1,2}$, the physical GDT curves are not well suited anymore, while for $\sigma_{G1,2}$ higher than 0.2 GHz the calculated GDT decreases (rises) too early in correspondence to the lower (higher) transitional band with respect to the observed real cases. From practical observations and from [26], the values τ_{m1} and τ_{m2} are assumed equal $\tau_{m1} = \tau_{m2} = \tau_m$.

To this composed Gaussian shape a random background ripple $\rho_\tau(f)$ is added in order to simulate the non-flatness of the GDT in the passband. Also in this case, it is

---

[3]The absolute value on the ripple has been inserted in order to have $A(f) < 1$, i.e. no amplification is taken into account.

[4]For more details, refer to the Appendix.

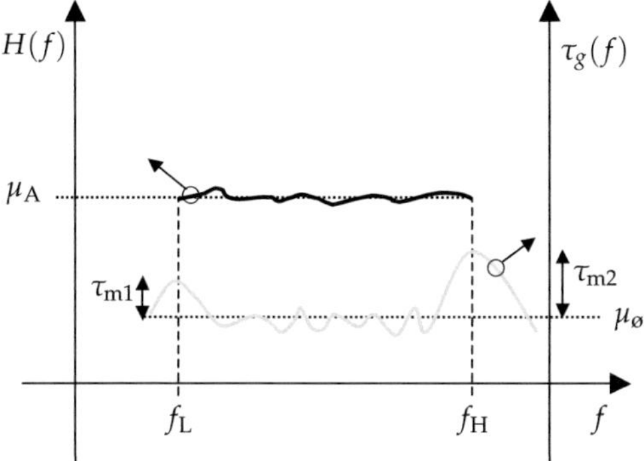

Figure 3.4: Illustration of the constructed non-ideal filter transfer function: the non-flat filter mask (black) and the non-constant GDT (gray) and some of the parameters defined in the previous sections.

assumed that the ripple is described by a Gaussian random process $\Xi_\tau(f)$ with mean μ_τ, standard deviation σ_τ, and correlation frequency separation $\Delta f_{\text{corr},\tau}$, namely

$$\rho_\tau(f) = \xi_i^\tau(f) \tag{3.8}$$

where ξ_i^τ is a generic realization of the process $\Xi_\tau(f)$.

The mean value μ_τ can be interpreted as the mean delay, which the signal undergoes when traveling through the filter structure. The mean GDT does not influence the shape of filter impulse response, it only shifts the filter impulse response. Hence, in the following analysis it is set to 0 ns. Supported by experimental results, in the EU-UWB case, the correlation frequency separation $\Delta f_{\text{corr},\tau}$ is set to 0.2 GHz and the standard deviation is taken as the parameter to be varied. Since only the in-band GDT is of interest, the overall GDT is given by

$$\tau_g(f) = \begin{cases} \gamma_\tau(f) + \rho_\tau(f) & f_L \leq f \leq f_H \\ 0 & f < f_L, f > f_H \end{cases} \tag{3.9}$$

The standard deviation of the GDT $\tau_g(f)$ is called σ_{GDT} and it is influenced by τ_m and σ_τ. In the following, τ_m is set to 1 ns (a typical value observed in measurements) while σ_τ is varied. The obtained random group delay is integrated over the frequency in order to obtain the phase $\angle H_{\text{rand}}$ of the filter transfer function, according to the GDT-phase relationship given by equation (2.3). Once the phase of the filter transfer function is calculated, it is inserted in equation (3.4).

An illustration of the random filter transfer function with both the random non ideal filter passband frequency mask and non constant GDT is shown in Fig. 3.4, together with some of the previously defined statistical parameters.

3.3.1.3 Monte Carlo Analysis

In order to quantify the influence of the ripple in the passband filter mask and of the GDT on the time domain filter performance, in a Monte Carlo analysis random transfer functions $H_{rand}(f)$ are calculated with a variation of both σ_A and σ_τ. From each random GDT realization, the value of σ_{GDT} is recovered. For each randomly calculated transfer function, the corresponding filter impulse response is calculated by equations (2.4)-(2.5). Then, the fidelity F is computed with equation (2.12) comparing the ideal filter impulse response to the randomly obtained one, together with P, τ_{FWHM}, and τ_r (ref. to eq. (2.7)-(2.9)). For each parameter pair (σ_A, σ_τ), 1000 realizations of the processes are generated with the following steps summarized in Tab. 3.1.

Table 3.1: Parameters' values of the statistical analysis

Parameter	Step
$0 \leq \sigma_A \leq 0.5$	0.025
$0 \leq \sigma_\tau \leq 1$ ns	0.05 ns

This also includes the case of a flat filter mask and non-constant GDT assuming $A(f) = 1$ for $f_L \leq f \leq f_H$ and varying only σ_τ. On the other hand, also the case of only a non-flat filter mask has been analyzed, setting $\tau_g = 0$ ns and varying only σ_A. The analysis of these two degenerated cases allows for evaluating the influence of each non-ideality itself.

3.4 Obtained Results

The results of the statistical analysis are reported and discussed in this section. Moreover, bounds on σ_A and σ_{GDT} that causes only negligible distortion are derived. In the analyzed case of UWB filters for the European regulation, the bandwidth is $B = 2.5$ GHz with $f_L = 6$ GHz and $f_H = 8.5$ GHz. With these assumptions the resulting values for the ideal filter are $\tau_{FWHM} = 0.48$ ns and $\tau_r = 1.06$ ns for $r = 0.1$. It has to be noticed that the filter impulse response has also in the ideal case a non-zero FWHM and ringing, due to the fact that $B < \infty$.

In the following figures 3.5 - 3.8, the mean value of each parameter (F, P, τ_{FWHM} and τ_r) over the 1000 realizations for each pair (σ_A, σ_{GDT}) is plotted. The points located on the horizontal axis, i.e. at the positions $(\sigma_{GDT}, 0)$, represent the case in which only a non-constant GDT is present (i.e. $A(f) = 1$ for $f_L \leq f \leq f_H$). The points located on the vertical axis, i.e. at the positions $(0, \sigma_A)$, represent the case of a constant GDT and non-flat filter mask. The ideal case (flat passband filter mask and constant GDT) is located in the plot at the position $(0,0)$.

3.4.1 Fidelity Analysis

First, the results of the fidelity analysis are regarded. As it can be seen from Fig. 3.5, F depends on both σ_A and σ_{GDT}, i.e., both non-idealities contribute to deteriorate the time domain filter behavior with respect to the ideal case. For low values of GDT variation and for low ripple in the passband frequency mask, the fidelity F is mainly affected by the GDT variation. As σ_A increases, also the passband mask ripple influences the fidelity F. Nevertheless, σ_{GDT} is much more critical than σ_A. The highest values of the fidelity ($F > 0.9$) result for $\sigma_A < 0.35$ and $\sigma_{GDT} < 0.1$ ns in a small region of the (σ_A, σ_{GDT}) plane. It is obvious that combinations of $\sigma_A < 0.35$ and $\sigma_{GDT} < 0.1$ ns should be the goal of UWB bandpass filter design in order to guarantee high fidelity of the output signal.

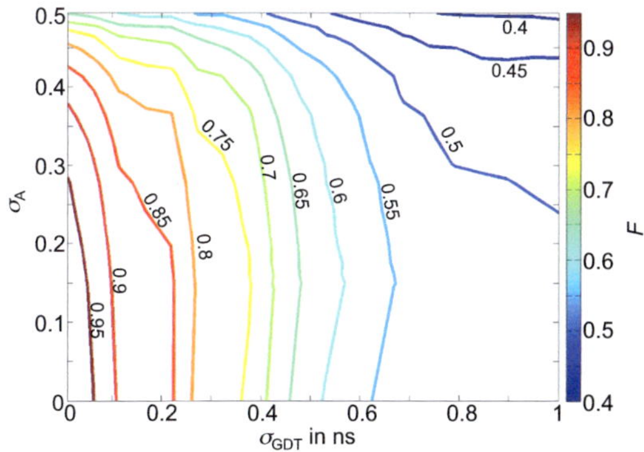

Figure 3.5: Fidelity F between the ideal and the random filter impulse responses for different values of σ_A and σ_{GDT}.

3.4.2 Peak P

In Fig. 3.6 the peak values P of the filter impulse responses are plotted. The plot is normalized to the peak of the ideal filter impulse response (hence in (0,0) $P = 1$). The peak values P depend strongly on both σ_A and σ_{GDT}, which cause the peak P to decrease as they increase. As discussed in section 3.2.1, the ripple in the filter GDT causes different time delays of the different frequency components. Consequently the pulse is spread over time and the peak P decreases. This spreading of the pulse in time can be also inferred looking at the previous Fig. 3.5, where the fidelity is very low for high value of σ_{GDT}. It can be derived that for $\sigma_A < 0.15$ and $\sigma_{GDT} < 0.25$ ns the normalized peak P is high ($P > 0.95$).

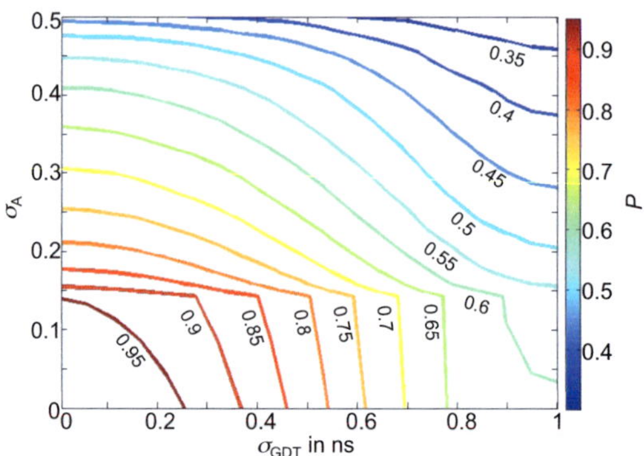

Figure 3.6: Peak P of the random filter impulse responses for different values of σ_A and σ_{GDT} normalized to the peak of the ideal filter impulse response.

3.4.3 FWHM τ_{FWHM}

In Fig. 3.7, the values of τ_{FWHM} of the filter impulse responses, calculated according to eq. (2.8), are illustrated. The minimum τ_{FWHM} that can be achieved for the EU-UWB case is 0.5 ns. Observing the plot, it can be inferred that the FWHM is mainly influenced by the ripple in the filter GDT, whereas the ripple in the passband filter mask has little influence on this parameter. On the other hand, the presence of a ripple in the GDT causes the different components to be delayed by different delays and hence the pulse is spread in time, as clarified in section 3.2.1. From Fig. 3.7 it can be evinced that a small increase (lower than 20%) of the filter FWHM with respect to the ideal case can be obtained for $\sigma_{GDT} < 0.1$ ns, while σ_A does not influence this value. In that case it is guarantee that $\tau_{FWHM} < 0.6$ ns.

3.4.4 Ringing τ_r

In Fig. 3.8 the ringing duration τ_r of the filter impulse response for $r = 0.1$ is plotted. The ringing is influenced by both the GDT and the ripple in the passband filter mask. From Fig. 3.8 it can be seen that a maximum increment of the ringing of approximately 45% ($\tau_r < 1.5$ ns) occurs only in a small area of the (σ_A, σ_{GDT}) plane for $\sigma_A < 0.2$ and $\sigma_{GDT} < 0.1$ ns. For any higher value of σ_A and σ_{GDT} the ringing increases drastically. Hence, both non-idealities have a very high impact on the ringing duration.

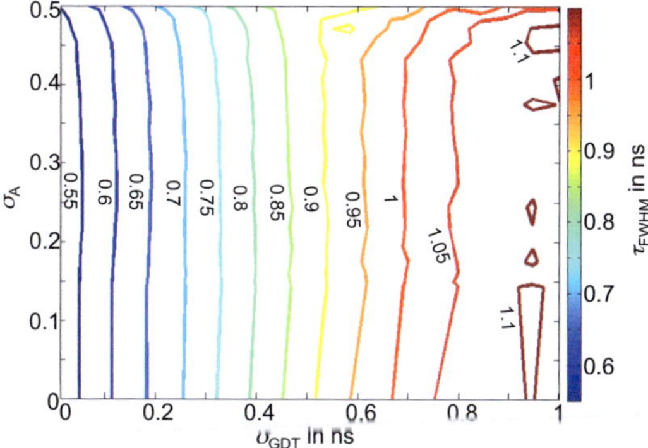

Figure 3.7: FWHM τ_{FWHM} of the random filter impulse responses for different values of σ_A and σ_{GDT}.

3.4.5 Derived bounds

From the performed analysis it can be seen that there exists bounds on the standard deviation of the passband mask and of the GDT that must be respected in order to have only negligible distortion compared to the ideal filter. These bounds are depending on the filter bandwidth and centre frequency. Since the previous simulations have been conducted for the European regulation, the bounds derived here will also apply only for this specific application. Nevertheless, by repeating the described Monte Carlo simulation procedure for another filter design goal (such as FCC regulation), similar bound values can be obtained for other regulations.

According to the results obtained with the performed analysis, in order to have a quasi-ideal time domain behavior of UWB filters for the EU-UWB regulation, the maximum allowed variation of the ripple in the passband filter mask is $\sigma_A < 0.15$ and for the GDT $\sigma_{\text{GDT}} < 0.1$ ns. With these values the decrement of the peak P of the filter impulse response with respect to the ideal case is only 5%, i.e. $P = 0.95$ relative to the peak of the ideal case, the FWHM has an increment lower than 20% (being $\tau_{\text{FWHM}} < 0.6$ ns), the ringing duration is lower than 1.5 ns (its increment with respect to the ideal case is lower than 45%). Moreover, the fidelity F is high (higher than 0.9). Hence, with these bounds the filter presents a good, quasi-ideal, time domain behavior and pulse preserving capability.

The validity of these bounds will be proven in the following with measurement results from filters developed for the EU-UWB regulation.

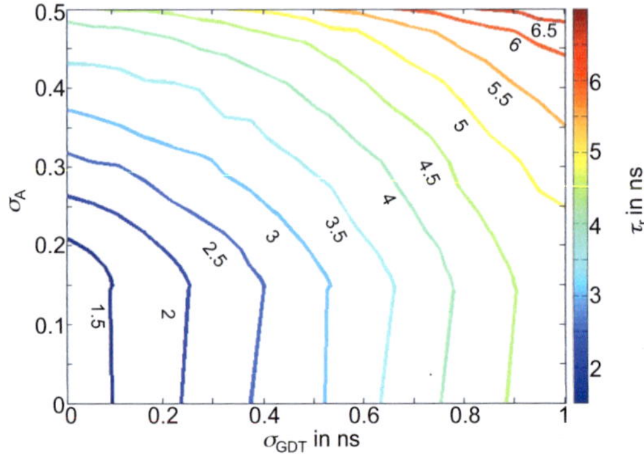

Figure 3.8: Ringing duration τ_r with $r = 0.1$ of the random filter impulse responses for different values of σ_A and σ_{GDT}.

3.5 Practical UWB Filter Typologies

In this section different approaches to the design of UWB filters are investigated and compared. Prototypes for different realization techniques are fabricated and characterized with classical frequency domain parameters as well as in the time domain. Of particular interest is the impact on the time domain filter behavior of the different realization typologies. Moreover, the previously introduced analysis is applied and validated through measurement results. Firstly, techniques for the realization of UWB bandpass filters are presented. Secondly, different UWB bandpass filter typologies are investigated by analyzing the measured results of fabricated filter prototypes.

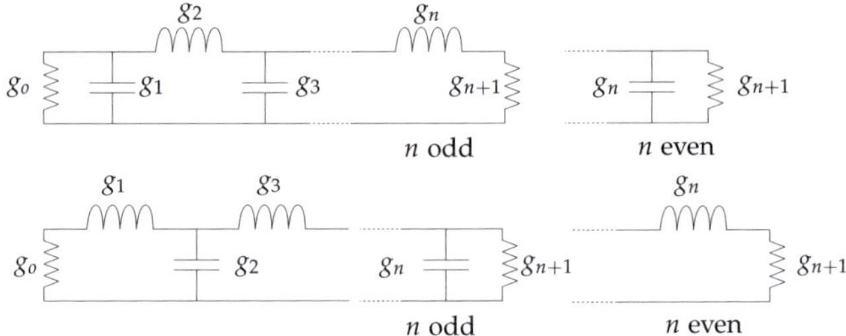

Figure 3.9: The lowpass filter model.

3.5.1 Filter Design

In order to realize the desired filter transfer function, a filter synthesis has to be performed. This is accomplished starting from a lowpass filter model composed of a cascade of inductors L_i and capacitors C_i, whose values g_i are normalized so that the filter input resistance (or conductance) is unitary and the filter has a unitary cut-off frequency [14], [15]. The schematic representation of this filter model is given in Fig. 3.9.

The first step in the design procedure consists in selecting the desired filter characteristic function $F_n(f)$ and filter order n that satisfy the given requirements. Then, the values of each element g_i can be calculated or derived from tabulated values.

Once the structure is defined, it has to be converted into the desired network topology (lowpass, bandpass, highpass, bandstop) and de-normalized to the desired input impedance Z_0. This conversion can be made using frequency transformations [14], [15].

Lowpass Transformation This kind of transformation permits to obtain practical lowpass filters, with desired cutoff frequency f_C and input impedance Z_0, starting from the previously described model.

Highpass Transformation It permits to derive an highpass filter with desired cut-off frequency f_C and input impedance Z_0, starting from the previously described lowpass model.

Bandpass Transformation With this transformation it is possible to obtain a bandpass filter with desired input impedance Z_0, lower cut-off frequency f_L and upper cut-off frequency f_H, starting from the previously described lowpass model.

Bandstop Transformation It permits to derive a bandstop filter with desired stop band and input impedance Z_0, starting from the previously described lowpass mode.

Richard's Transformation These particular transformations permit to convert lumped elements into distributed transmission line elements for practical implementation of microwave filters.

Once the de-normalized filter structure has been obtained, i.e. the actual values of C_i' and L_i' are calculated, the impedance of the connecting lines corresponding to the values required for the lumped elements is derived.

3.5.2 UWB Filter Typologies

The task of realizing filters operating in the Ultra Wideband range is quite challenging. In literature, different filter structures have been proposed using different realization techniques. One classical technique consists in realizing filters with resonators at the center frequency of the UWB band [16], [17], [18]. This technique is relatively easy to implement and it is consolidated with a rather large number of references [14],

[15], [44]. In the case of UWB filters, classical methods like end-coupled or parallel-coupled techniques are usually not implemented since the gaps between the coupled lines required to obtain such a huge bandwidth are often too small and hence not realizable.

Other methods consist in realizing UWB bandpass filters through a cascade of a lowpass and a highpass section [24], [49] and in some cases these two sections are integrated into the same unit to decrease the overall filter dimensions. Recently, some contributions have proposed UWB bandpass filters using back-to-back line transitions [50], [51], [52], which can be implemented as microstrip-to-slotline or microstrip-to-CPW transitions.

In the following, these most frequently used and important filter typologies are investigated in order to find out the characteristics of each particular realization technique. For each approach, a prototype filter is designed, fabricated and characterized in the frequency domain and in the time domain. The previously developed analysis based on the standard deviations of the GDT and of the passband ripple is here used in order to quantify the filter time domain behavior and pulse preserving capability. Moreover, the random analysis is validated through measurement results.

3.5.3 Band Pass Filters with Resonator Stubs

This filter typology permits to obtain a bandpass behavior using stub resonators, whose length is a quarter-wavelength at the center frequency of the bandpass interval. This typology permits to realize different filter masks (i.e. Chebychev, Butterworth, ...) and can be implemented using two main methods: short-circuited stubs realizations and open-end stubs realizations (see Fig. 3.10) [14], [15].

Short-Circuited Stubs Filters The transmission line model of this filter implementation is given in Fig. 3.10(a). It consists of n short-circuited stubs with length $\ell_S = \lambda_{g,c}/4$, connected by $n-1$ lines with length $\ell_C = \lambda_{g,c}/4$, where $\lambda_{g,c}$ represents the guided wavelength[5] at the center frequency f_c and n the filter order.

Open-End Stubs Filters The open-end method (Fig. 3.10(b)) is composed by n open-end stubs with length $\ell_S = \lambda_{g,c}/2$, connected by $n-1$ lines with length $\ell_C = \lambda_{g,c}/4$. This structure possesses additional passbands also in the vicinity of $f = 0$ and at the frequencies $f = 2kf_c$ with k integer. The open-end stubs are composed of two sections with different lengths and impedances. By changing this length it is possible to change the location of the poles with respect to the previous method and hence to change the transitional band/stopband characteristic.

Pseudo-Highpass Filters It has to be included into this section of resonator filters also another class of bandpass filters which are derived from highpass structures.

---

[5]For a substrate with permittivity ε_r, the guided wavelength $\lambda_{g,c}$ corresponding to the in-vacuum wavelength λ_c is given by

$$\lambda_{g,c} = \frac{\lambda_c}{\sqrt{\varepsilon_r}}. \tag{3.10}$$

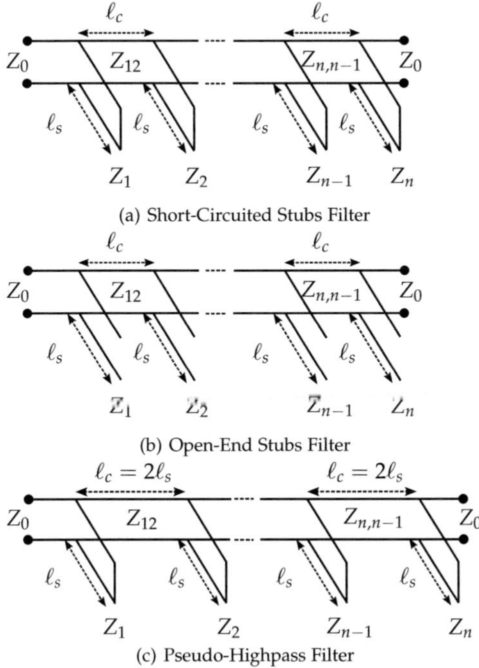

(a) Short-Circuited Stubs Filter

(b) Open-End Stubs Filter

(c) Pseudo-Highpass Filter

Figure 3.10: Realization schemes of bandpass filters with resonator stubs.

This kind of filters are based on the periodic structure of high pass filters [15]. They are composed of n short circuited stubs of length $\lambda_{g,L}/4$ connected by lines of length $\lambda_{g,L}/2$, where $\lambda_{g,L}$ is the guided wavelength corresponding to the lower cut-off frequency f_L (see Fig. 3.10(c)). A filter order n of this filter typology is equivalent to a filter order $2n - 1$ of a conventional band pass filter [15].

The value of the stub impedances Z_i and of the connecting-line impedances Z_{ij} can be directly obtained from tabulated filter parameters, according to the desired filter mask and order.

In order to analyze the frequency and time domain behavior of bandpass resonator filters, different prototypes have been fabricated and tested. In the following, two examples are shown, and the main characteristics of this filter typology are pointed out.

- *Model 1*: The fabricated filter, implementing a 7th-order bandpass Chebychev mask, has been etched on a Rogers substrate RO3206, with $\varepsilon_r = 6.15$ and thickness $h = 0.64$ mm (see Fig. 3.11).

- *Model 2*: The fabricated filter, implementing a 6th-order bandpass Chebychev mask, has been etched on a Rogers substrate RO4003, with $\varepsilon_r = 3.38$ and thickness $h = 1.57$ mm (see Fig. 3.12).

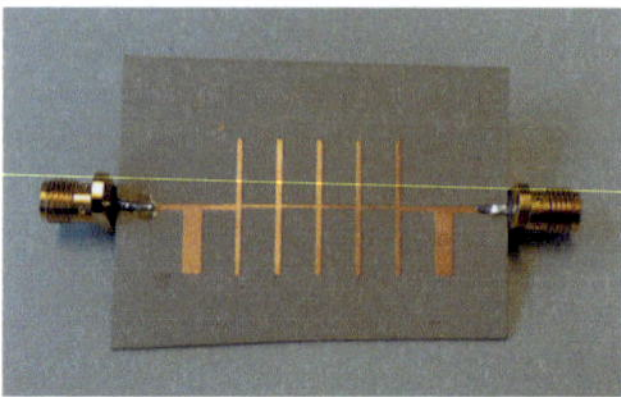

Figure 3.11: The realized filter structure: Model 1.

Figure 3.12: The realized filter structure: Model 2.

3.5.3.1 Model 1: Frequency domain analysis

The fabricated prototype of the filter structure Model 1 has been measured in the frequency domain with a VNA. The measurement has been performed in the frequency interval $2 - 18$ GHz with a resolution of $N = 801$ points. The frequency resolution of the acquired data is $\Delta f = (f_H - f_L)/N = 20$ MHz. The absolute values of the measured S_{21} and S_{11} parameters are shown in Fig. 3.13. The measured filter mask ($|S_{21}|$ parameter) perfectly matches with the design goal of the filter mask. The width of the transitional bands is very small. In fact, as previously said, the filter is based on a 7th order Chebychev structure. Due to this high filter order, the realized structure presents sharp transitional bands. The passband ripple of the filter mask is high (3 dB). This is mainly due to the selected stub realization and of the selected Chebychev characteristic function, which has the advantage of having sharp transitional bands but it oscillates in the passband [15]. In Fig. 3.14, the GDT parameter is shown. It has been calculated starting from the measured phase of the S_{21} parameter and then applying to it a derivation with respect to the frequency, according to eq. (2.3). The GDT presents two peaks which can be approximated by a composition of two Gaussian functions (see Fig. 3.14): one at the lower cut-off frequency, 6 GHz, and the second at the upper cut-off frequency, 8.5 GHz. The GDT is non-constant in the passband, but presents low oscillations. These oscillations are due to energy storage and resonances inside the structure itself. The transition regions from 5.5 GHz to 6 GHz and 8.5 GHz to 9 GHz cause clearly visible distortions in the group delay at these frequencies. The measured filter GDT perfectly matches with the assumption of the developed random model of the GDT.

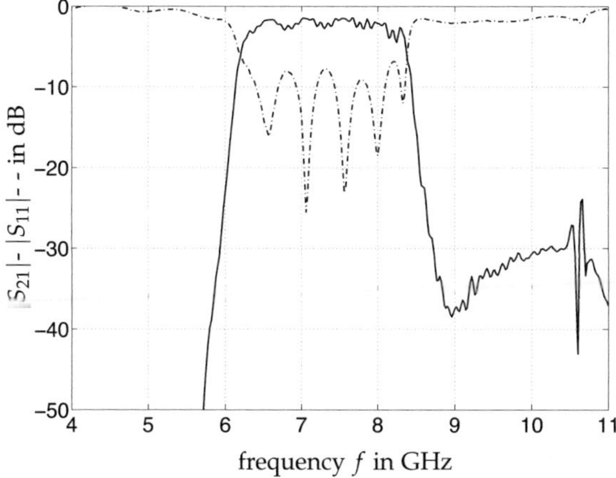

Figure 3.13: Measured $|S_{21}|$ (solid line) and $|S_{11}|$ (dotted line) parameters of the fabricated resonator stub filter Model 1.

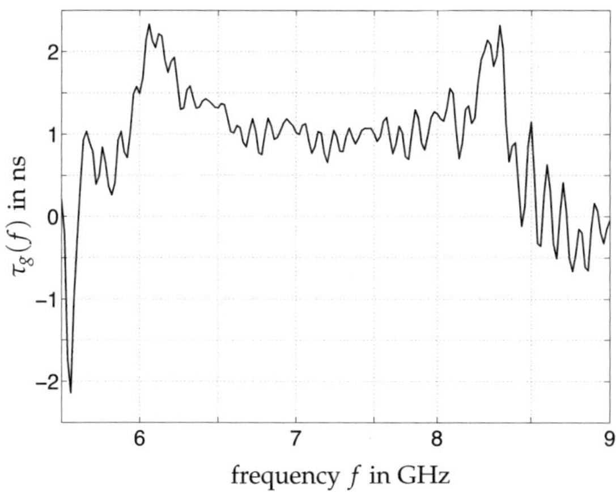

Figure 3.14: Passband zoom of the measured GDT $\tau_g(f)$ of the fabricated resonator stub filter Model 1.

In Tab. 3.2 the measured values of the frequency domain non-idealities for this filter are summarized.

Table 3.2: Measured value for the developed analysis (Model 1).

Parameter	measured value
σ_A	0.393
σ_{GDT}	0.3 ns

3.5.3.2 Model 2: Frequency domain analysis

The same measurement procedure of filter Model 1 has been applied to filter Model 2. In Fig. 3.15 the absolute values of the measured S_{21} and S_{11} parameters are illustrated. This filter typology presents a good passband and stopband behavior, with a low level of passband ripple and a high stopband attenuation. The transition bands are quite sharp, due to the high filter order and the characteristics of the Chebychev mask. In Fig. 3.16, the GDT parameter is illustrated. It presents low oscillation in the pass-band. The transition regions from 5 GHz to 6 GHz and 8.5 GHz to 9.5 GHz cause clearly visible distortions in the group delay at these frequencies. Also in this case the GDT presents two peaks that can be approximated by Gaussian functions (see Fig. 3.17): one at the lower cut-off frequency, 6 GHz, and the second at the upper cut-off frequency, 8.5 GHz.

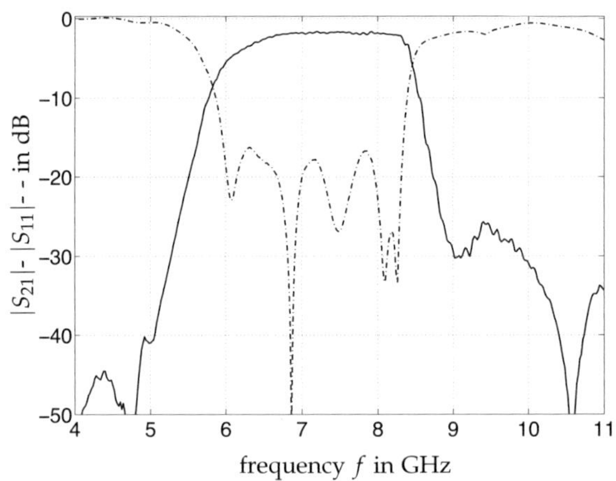

Figure 3.15: Measured $|S_{21}|$ (solid line) and $|S_{11}|$ (dotted line) parameters of the fabricated resonator stub filter Model 2.

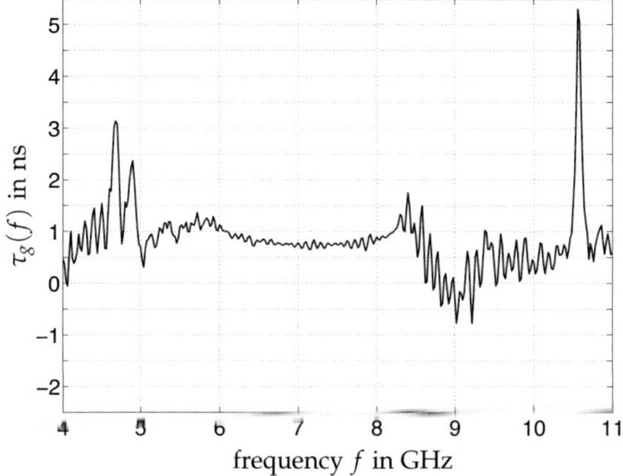

Figure 3.16: Measured GDT $\tau_g(f)$ of the fabricated resonator stub filter Model 2.

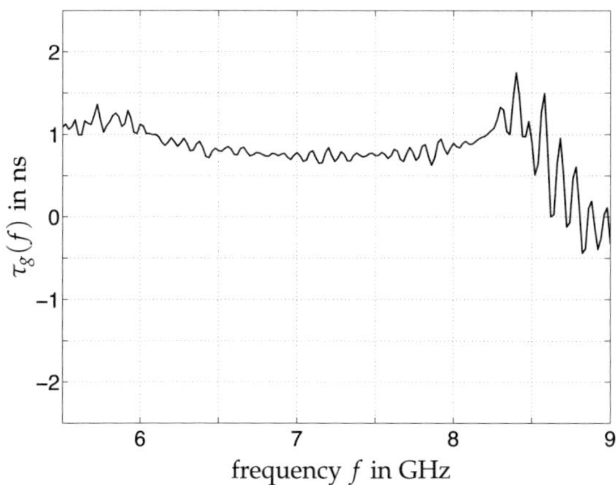

Figure 3.17: Zoom of the measured GDT $\tau_g(f)$ of the realized resonator stub filter Model 2 in the passband.

In Tab. 3.3 the measured values of the frequency domain non-idealities for this filter are summarized.

Table 3.3: Measured value for the developed analysis (Model 2).

Parameter	measured value
σ_A	0.22
σ_{GDT}	0.143 ns

3.5.3.3 Model 1: Time domain analysis

The filter impulse response has been calculated according to the method illustrated in section 2.2. The recovery of the time domain data has been performed as described in the Appendix in section A.1. Starting from the measured filter transfer function $H(f) = S_{21}(f)$, zeros have been inserted between 0 and 2 GHz and a zero-padding at the end of the sample vector from 18 to 50 GHz has been performed, in order to increase the time domain accuracy. The analytic filter transfer function $H^+(f)$ has been constructed as (see eq. (2.4))

$$H^+(f) = \begin{cases} 2H(f) & f > 0 \\ H(f) & f = 0 \\ 0 & f < 0 \end{cases} \tag{3.11}$$

From the analytic transfer function the filter impulse response has been recovered via IDFT, namely (see eq. (A.1))

$$h(k\Delta t) = \Re \left[\sum_{n=0}^{\tilde{N}-1} H^+(n\Delta f) \cdot \exp\left[j\frac{2\pi}{\tilde{N}} kn \right] \right] \tag{3.12}$$

where $\tilde{N}$ is the number of points resulting after the zero padding.

Together with the filter impulse response, also the impulse response of the ideal filter has been calculated. It has been obtained by generating a mask with a constant value of 1 in the passband ($6 - 8.5$ GHz) and zeros elsewhere in the frequency interval $0 - 50$ GHz, with a frequency resolution of $\Delta f = 20$ MHz, as the measured data. From its analytic transfer function, the impulse response has been recovered via IDFT.

The impulse response of the realized filter has been normalized to the maximum value of the envelope of the impulse response of the ideal filter. This permits to directly assess the impact of the non-ideal filter behavior with respect to the ideal filter.

The impulse response of the filter Model 1 is illustrated in Fig. 3.18 together with its envelope. It has been normalized to the peak of the ideal filter impulse response. Its normalized peak is $P = 0.651$, i.e. it has a decrement of 35.5% with respect to the ideal case. The FWHM is $\tau_{FWHM} = 0.61$ ns and the ringing is (for $r = 10\%$) $\tau_{0.1} = 1.92$ ns. In Tab. 3.4 the measured value of the filter time domain parameters are compared with the predicted values from the developed statistical analysis. The values predicted from the statistical analysis has been obtained looking at the calculated 2D plots (see Fig. 3.6, 3.7 and 3.8) for the pair $(\sigma_A, \sigma_{GDT}) = (0.393$ ns, 0.3 ns$)$ (see Tab. 3.2).

The time domain parameters calculated after the measurements are in accordance with the prediction of the developed statistical analysis.

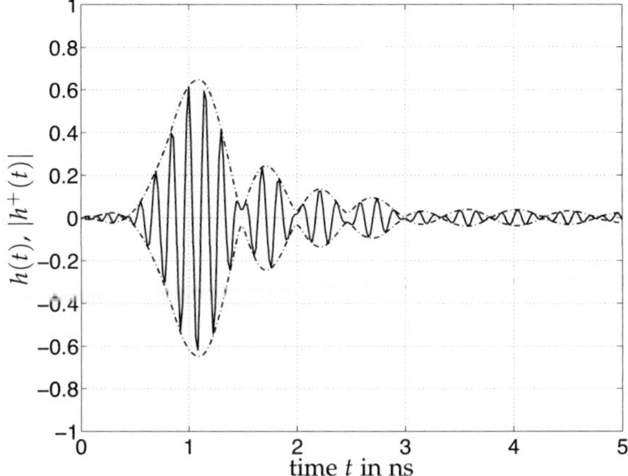

Figure 3.18: Measured impulse response $h(t)$ (solid line) and its envelope $|h^+(t)|$ (dotted line) of the fabricated resonator stub filter Model 1, normalized to the peak of the ideal filter impulse response peak.

Table 3.4: Time domain parameters of the resonator stub filter Model 1: measured values vs. values predicted from statistical analysis

Parameter	measured value	predicted value
P	0.651	0.7
τ_{FWHM}	0.61 ns	0.65 ns
$\tau_{0.1}$	1.92 ns	2.2 ns

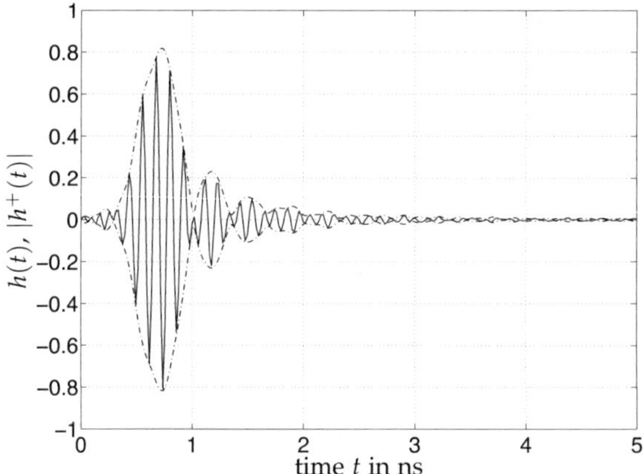

Figure 3.19: Measured impulse response $h(t)$ (solid line) and its envelope $|h^+(t)|$ (dotted line) of the fabricated resonator stub filter Model 2, normalized to the peak of the ideal filter impulse response peak.

3.5.3.4 Model 2: Time domain analysis

In Fig. 3.19 the impulse response of the filter Model 2, together with its envelope, is illustrated, normalized to the peak of the ideal filter impulse response. The normalized peak is $P = 0.8133$, i.e. it has a decrement of 18.7% with respect to the ideal case. The FWHM is $\tau_{\text{FWHM}} = 0.48$ ns and the ringing are 1.02 ns.

In Tab. 3.5 the measured time domain parameters together with those predicted from the performed statistical analysis are given.

Table 3.5: Time domain parameters of the resonator stub filter Model 2: measured values vs. values predicted from statistical analysis

Parameter	measured value	predicted value
P	0.8133	0.85
τ_{FWHM}	0.48 ns	0.5 ns
$\tau_{0.1}$	1.02 ns	2.2 ns

The time domain parameters are in accordance with the prediction of the statistical analysis.

It has to be pointed out that the ringing effect of this filter structure is mainly caused by energy storage, which is responsible for the dispersion of the signal in time. In fact, this kind of filter is realized using quarter-wavelength resonators. Hence, it can be concluded that the resonance effect degrades the time domain behavior of the filter, causing long ringing duration in the impulse response.

3.5.4 Integrated lowpass-highpass sections filters based on microstrip-to-slotline transitions

In this section another typology of UWB bandpass filters is analyzed. This kind of filter consists in the integration of a lowpass and a highpass section. The integration of these two sections permits to reduce the used space with respect to a classical cascade of two sections. The filter here regarded is realized as a back-to-back structure, which is realized through microstrip-to-slotline transitions.

The highpass behavior is obtained through microstrip-to-slotline transitions. These transitions are formed by the series of two T-junctions; at the input by a microstrip-to-slotline transition and at the output by a slotline-to-microstrip transition. For an efficient wideband coupling from the slotline to the microstrip line, circular cavities are inserted as inductive elements at both ends of the slotline, and circular patches are inserted at the terminations of the microstrip lines [53]. The low-pass unit is composed by resonators realized on the slotline, which exhibit a low radiation loss. These stubs are dimensioned in order to have approximately a length of $\ell_S = \lambda_g/4$ (where λ_g is the guided wavelength) at the upper frequency of the EU-UWB band $f_H = 8.5$ GHz and are in series connected by lines with length $\ell_C = \lambda_g/8$.

A filter prototype etched on a substrate Rogers RO4003 with $\varepsilon_r = 3.38$ and thickness $h = 0.51$ mm (Fig. 3.20) has been fabricated and measured. Its dimensions are very small (substrate dimension: 2.5×3 cm^2; filter dimension 12×6 mm^2).

Figure 3.20: The fabricated filter with integrated lowpass-highpass sections: rear side (left) and front side (right).

3.5.4.1 Frequency domain analysis

The absolute values of the measured S_{21} and S_{11} parameters are illustrated in Fig. 3.21. The $|S_{11}|$ is below -18 dB in the whole filter pass-band. The upper transition band is steeper compared to the lower transition band, which is mainly due to the microstrip-to-stripline transition [53]. The filter mask presents a small passband ripple whose standard deviation to the mean value in the filter pass-band is 0.58 dB. Also this effect is mainly a result of the microstrip-to-slotline transition, as described in [53].

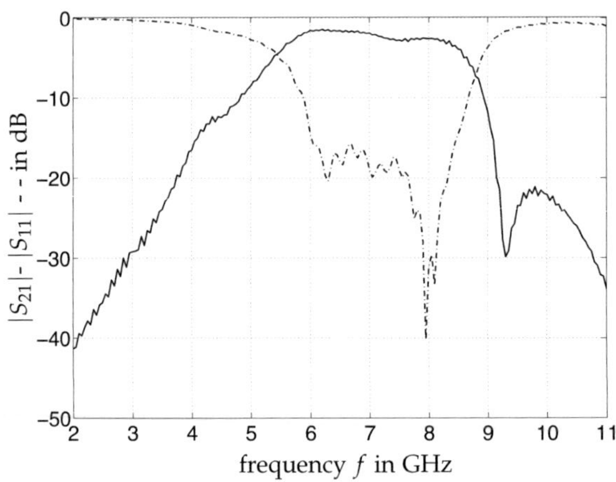

Figure 3.21: Measured $|S_{21}|$ (solid line) and $|S_{11}|$ (dotted line) parameters of the fabricated filter.

The group delay time (Fig. 3.22) is very flat in the filter passband. Its mean value is 0.3729 ns and it has a standard deviation in the passband of 58.1 ps.

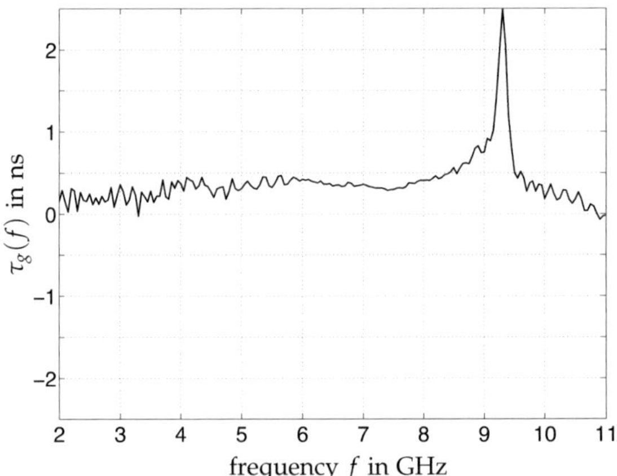

Figure 3.22: Measured GDT $\tau_g(f)$ of the fabricated filter.

In Tab. 3.6 the measured standard deviations of the frequency domain non-idealities for this filter are summarized.

Table 3.6: Measured value for the developed analysis.

Parameter	measured value
σ_A	0.23
σ_{GDT}	0.0581 ns

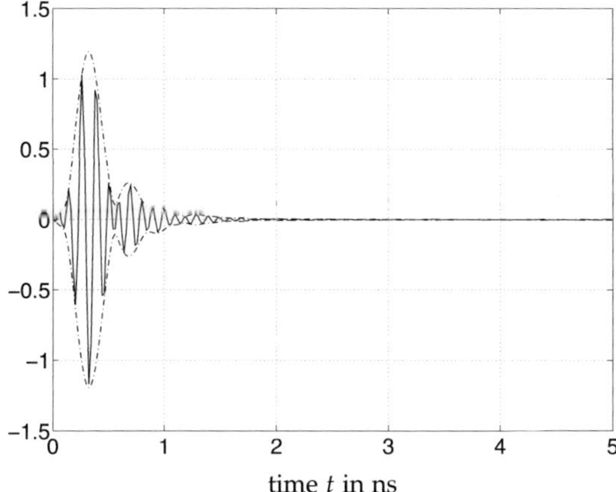

time t in ns

Figure 3.23: Measured impulse response $h(t)$ (solid line) and its envelope $|h^+(t)|$ (dotted line) of the fabricated transition filter, normalized to the peak of the ideal filter impulse response peak.

3.5.4.2 Time domain analysis

The filter transient response has been calculated using the method illustrated in section A.1, as for the resonator stub filters. The filter impulse response obtained for the transition filter is illustrated in Fig. 3.23 together with its envelope. The plot has been normalized to the peak of the impulse response envelope of the ideal filter. The peak value is high: its normalized value is 1.16. This is a result of the smooth transition of the highpass section, which allows for a considerable amount of power below 6 GHz to pass the filter, which then contributes to increase the peak of the filter impulse response. The FWHM is small ($\tau_{\mathrm{FWHM}} = 0.4$ ns) and the ringing duration is quite low. The ringing duration for a value of the parameter $r = 10\%$ is $\tau_{0.1} = 0.66$ ns. The presence of ringing is mainly resulting not only from the non-perfect constant group delay time, but also from the limited flatness of the filter pass-band mask.

In Tab. 3.7 the calculated time domain parameters of the filter are given, together with the predicted values obtained from the proposed statistical analysis for the measured pair $(\sigma_A, \sigma_{\mathrm{GDT}}) = (0.23, 0.0581 \text{ ns})$.

In comparison with the stub realization, the transition filter offers better time domain performance, but at the same time the sharpness of the transition band is limited due to the characteristics of the microstrip-to-slotline transition.

Table 3.7: Time domain parameters of the transition filter: measured values vs. values predicted from statistical analysis.

Parameter	measured value	predicted value
P	1.16	0.8
τ_{FWHM}	0.4 ns	0.5 ns
$\tau_{0.1}$	0.66 ns	1.15 ns

3.5.5 CPW Filters

In this section a third typology of UWB bandpass filters is investigated: filters realized through Coplanar Wave Guide (CPW) technology. In the following, the basic elements for a filter realization in CPW technology are regarded. Then, the realized filter prototype is presented.

3.5.5.1 CPW Filter Elements

The basic elements in CPW technology are realized through discontinuities in the CPW structure itself [54], [55], [56]. The CPW structure without discontinuities is shown in Fig. 3.24(a). The most important discontinuities, which represents fundamental components for the realization of filters, are now summarized.

Step Change in Width A step change in the width of the CPW center conductor (see Fig. 3.24(b)) can be modeled as a shunt capacitance C_L. It has a lowpass behavior.

Series Gap A series gap in the CPW center conductor can be modeled as a lumped π-network (see Fig. 3.24(c)) consisting of a coupling series capacitance C_{g_2} and two fringing shunt capacitances C_{g_1} and C_{g_3}, all of them being a function of the gap size.

Defected Ground Plane Slots etched in the CPW ground plane, as shown in Fig. 3.24(d), act as bandstop filters, whose center frequency is directly proportional to the slot length. In first approximation, they can be modeled as a series inductance.

Open-End Series Stub An open-end series stub realized in CPW technology is illustrated in Fig. 3.24(e). It can be modeled, in first approximation, as a series capacitance C_o, whose value is function of the finger dimensions. This element presents a highpass behavior.

Short-End Series Stub A short-end CPW series stub (see Fig. 3.24(f)) can be modeled, in first approximation, as a series inductance and presents a bandstop behavior.

Open-End Shunt Stub An open-end CPW shunt stub is shown in Fig. 3.24(g) (left). Air bridges have to be added in order to connect the ground for suppression of the coupled slotline mode. This structure can be modelled with the circuit shown in Fig. 3.24(g) (right), where the reactance X_s is due to the coplanar waveguide mode excited in the CPW stub and the reactance X_c is due to the coupled slotline mode, as described in [56]. C_a and L_a represent parasitic capacitance and inductance, respectively, resulting from the air bridges [55]. The complete structure has a lowpass behavior.

Short-End Shunt Stub A short-end CPW shunt stub (see Fig. 3.24(g), center) can be modeled, analogously to the open-end shunt stub. Also in this case air bridges have to be added for mode suppression. This has a highpass behavior.

Through an appropriate combination of these basic elements, different filter typologies (lowpass, highpass, ...) can be realized [57], [58].

3.5.5.2 CPW Bandpass Filter

Starting from these basic elements, a particular UWB filter structure has been realized. The filter is constructed by a cascade of a highpass section and a lowpass section. The highpass section is composed of two basic elements: the open-end series stub and the short-end shunt stub. The usage of two different elements for realizing the highpass section has been chosen in order to increase the sharpness of the transitional band at the cutoff frequency without highly increasing the occupied space, since they can be integrated one into the other. Also two element of only one typology could be used, but in this case, since elements of the same type cannot be easily integrated one into each other, the space occupancy would increase. The two selected different elements for the highpass section permit to be integrated one into the other and consequently to save space obtaining also the desired sharpness. The length of the shunt stub and of the open-end series stub has been set to a quarter wavelength corresponding to $f_L = 6$ GHz.

The lowpass section has been realized through a step change in the width of the central conductor. In order to highly suppress the upper band and make the transitional band sharper, instead of repeating the previous section, additional slots have been added in the CPW ground plane at both sides, realizing a defected ground plane structure. The length of these slots has been set to the guided quarter wavelength $\lambda_{gH}/4$ corresponding to the upper cut-off frequency $f_H = 8.5$ GHz. The complete lowpass section is composed of three step changes (each one of length $\lambda_{gH}/4$) and three symmetric slots in the CPW ground plane.

The final filter design consists of a cascade of two highpass sections and the described lowpass unit comprising three sections. A prototype of the developed filter structure has been etched on a substrate RO4003 with $\varepsilon_r = 3.38$ and thickness $h = 1.57$ mm and measured. The inner dimensions of the filter are 30×10 mm^2 (see Fig. 3.25).

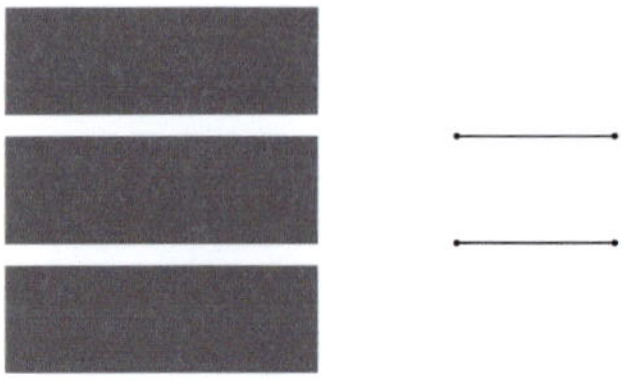

(a) CPW line

(b) Step Change in Width

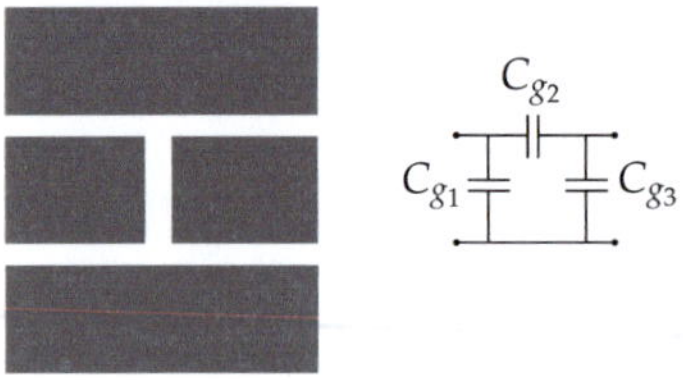

(c) Series Gap

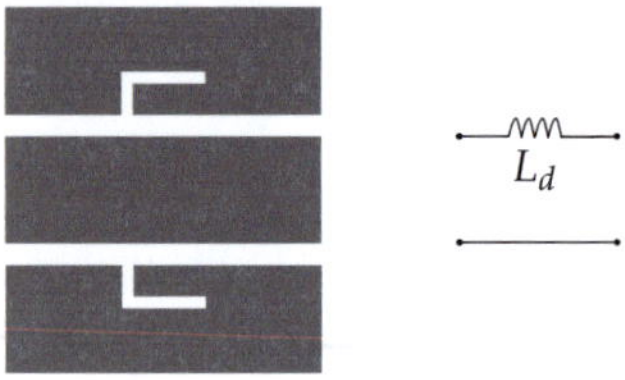

(d) Defected Ground Plane

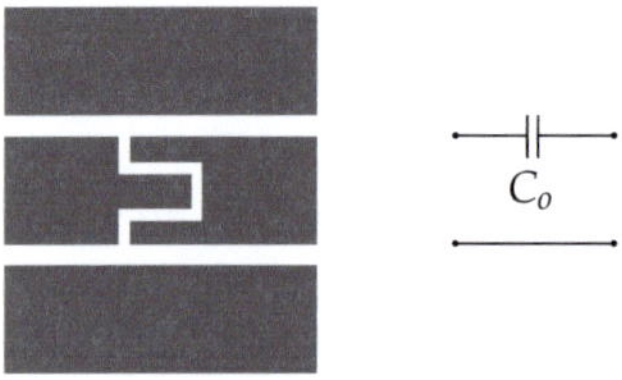

(e) Open End Series Stub

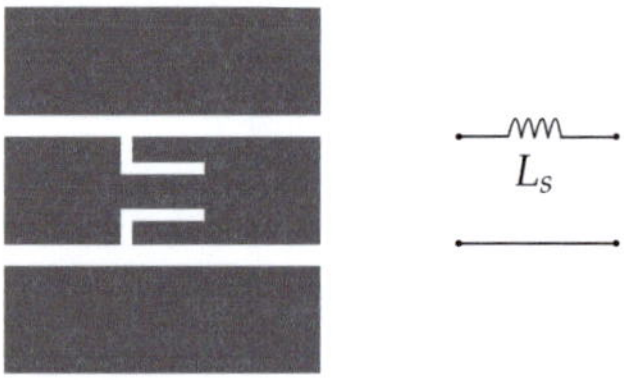

(f) Short End Series Stub

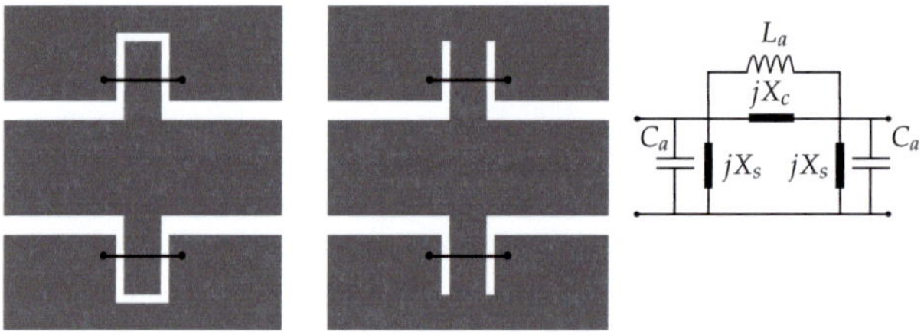

(g) Open End and Short Circuited Shunt Stubs

Figure 3.24: CPW Filter Elements.

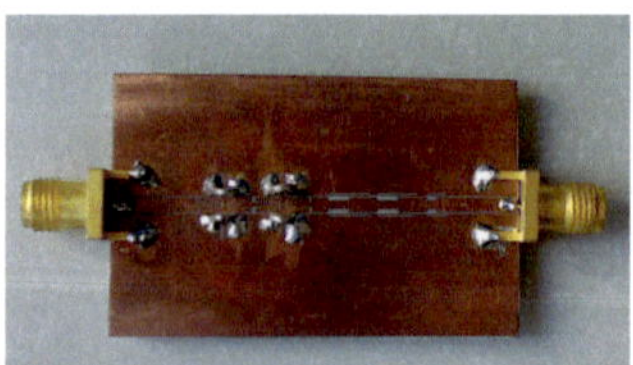

Figure 3.25: The realized CPW filter structure.

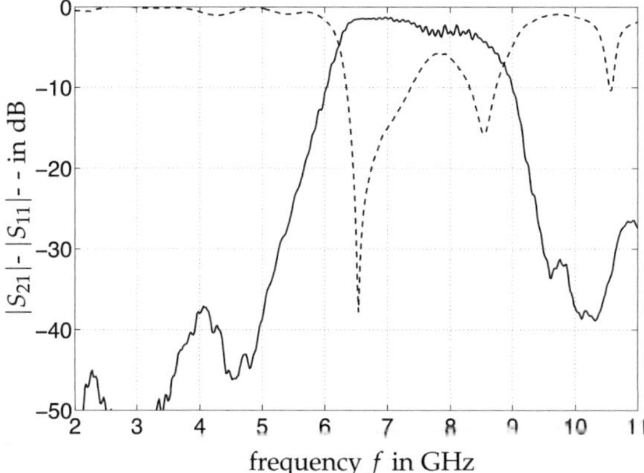

Figure 3.26: Measured $|S_{21}|$ (solid line) and $|S_{11}|$ (dotted line) parameters of the fabricated CPW filter.

3.5.5.3 Frequency domain analysis

The realized prototype has been tested with the same procedure as applied to the previous filters. The absolute values of the measured S_{21} and S_{11} parameters are shown in Fig. 3.26.

The realized filter has sharp transitional bands, as expected, due to the integration of multiple elements in the highpass section and in the lowpass section.

From the phase of the S_{21} parameter, also the GDT has been calculated. The result is plotted in Fig. 3.27.

The GDT presents oscillations in the stopband, while in the passband its variation is quite low. It has an in-band standard deviation of 0.23 ns.

The measured values of the standard deviations of the frequency domain non-idealities of the realized CPW filter prototype are summarized in Tab. 3.8.

Table 3.8: Measured values for the developed statistical analysis.

Parameter	measured value
σ_A	0.18
σ_{GDT}	0.143 ns

3.5.5.4 Time domain analysis

Together with a frequency domain analysis, also a time domain analysis has been performed.

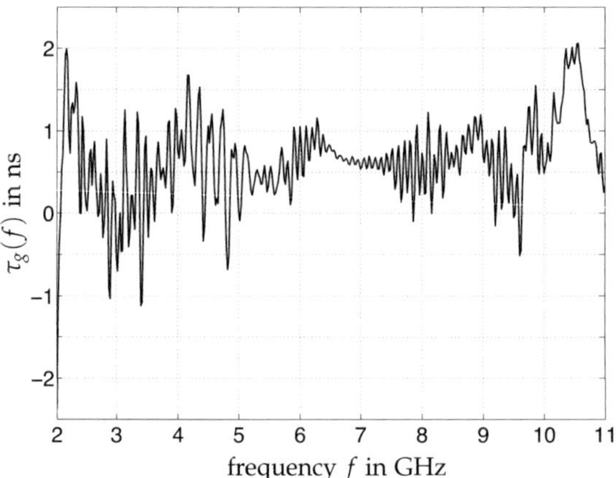

Figure 3.27: Measured GDT $\tau_g(f)$ of the realized CPW filter.

The calculated filter impulse response, using the same procedure as for the other filters, is plotted in Fig. 3.28, normalized to the peak of the ideal filter impulse response.

The obtained filter impulse response has a high relative peak P of 0.8521. The FWHM is small ($\tau_{\text{FWHM}} = 0.5$ ns) and the ripple level is also quite low ($\tau_{0.1} = 1.15$ ns).

The calculated time domain parameters are summarized in Tab. 3.9, compared with the results predicted from the developed statistical analysis for the measured pair $(\sigma_A, \sigma_{\text{GDT}}) = (0.18, 0.143\text{ns})$.

Table 3.9: Time domain parameters of the CPW filter: measured values vs. values predicted from statistical analysis

Parameter	measured value	predicted value
P	0.8521	0.85
τ_{FWHM}	0.5 ns	0.5 ns
$\tau_{0.1}$	1.15 ns	1.5 ns

3.6 Conclusion

In this chapter the time domain behavior of UWB bandpass filters has been investigated. In particular, the impact of the filter non-idealities in the frequency domain (non-flat filter frequency mask and non-constant filter group delay time) on the filter time domain behavior has been quantified. This investigation has been performed by

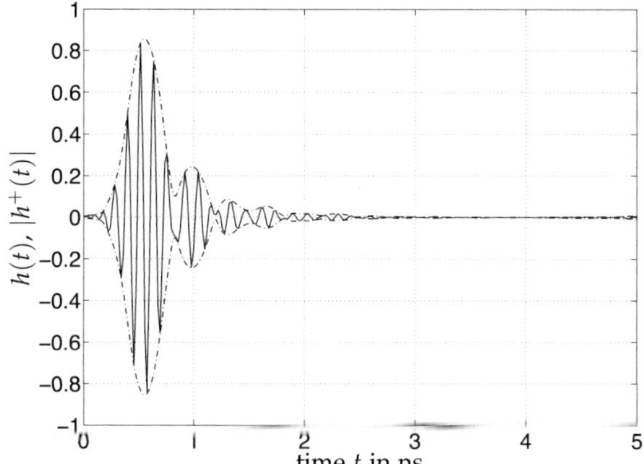

Figure 3.28: Measured impulse response $h(t)$ (solid line) and its envelope $|h^+(t)|$ (dotted line) of the fabricated CPW filter, normalized to the peak of the ideal filter impulse response.

a statistical analysis, as there is no deterministic solution, constructing filter transfer functions with a random ripple in the passband mask and a random ripple in the GDT. For that purpose suitable random distributions have been selected and models have been developed. By a Monte Carlo analysis, varying the ripple in the passband mask and the ripple of the GDT, different random filter transfer functions have been constructed and the related impulse responses have been calculated. Each obtained impulse response has been compared with the impulse response of the ideal filter. In order to quantify the variation from the ideal case, a correlation analysis between the random filter impulse response and the ideal impulse response has been conducted. Together with this analysis, also the time domain characterization parameters have been calculated, since the correlation analysis alone gives no information on the signal power.

From the performed investigation, it has been possible to define bounds on the standard deviations of the filter non-idealities in the frequency domain in order to have a quasi-ideal behavior in the time domain. Hence, it has been possible to directly assess the impact of the filter frequency domain non-idealities on its time domain behavior.

The novelty of the statistical approach presented in this chapter lies on the possibility to directly quantify the impact of the frequency domain non-idealities on the filter time domain behavior by regarding only the standard deviations of the filter passband mask and the GDT. The calculated graphs in Fig. 3.6-3.8 allow a quick evaluation of the time domain performance of UWB filters from measured frequency domain characteristics.

In the second part of the chapter different approaches to practical hardware realizations of UWB filters have been investigated. These realizations include classical approaches based on stub resonators as well as microstrip-to-slotline transitions and coplanar waveguide realizations. The characteristics of these design approaches have been compared in the frequency domain and in the time domain. It has been seen that the CPW filter realization offers the best performance by providing sharp transitional bands and good time domain behavior at the same time.

For each typology filter prototypes have been fabricated and tested. The proposed statistical analysis has been applied to these filters and validated.

4 Integration of UWB Antennas and Filters

Antennas are a key element of simple realizations of impulse radio UWB transmitters. However, due to regulation limitations, the spectral power density of the radiated electric field has to satisfy particular masks. For example, in the European context, the UWB signal has to respect the mask shown in Fig. 1.1. Moreover, in order to avoid interference with WLAN and other devices operating in particular bands, an UWB device should not radiate in particular frequency intervals like, e.g., the 5.8 GHz ISM band [59]. Also interference from narrow-band devices into the UWB receivers must be suppressed. Because of these problems (regulations, interference avoidance, etc.) and some related issues (such as an optimal utilization of the permitted transmit power, which is the topic of chapter 6), the spectrum of the radiated signal has to be properly shaped. A method to do this is to pre-filter the signal so that it does not possess spectral components in particular frequency bands. From this point of view, filters become important elements at the transmitter and the receiver side. Filters can be added in cascade to the antenna, or they can be directly integrated in the antenna board.

The topic of this chapter is hence the integration of filters and UWB antennas. The chapter is organized as follows. Firstly, the investigation in the time domain and in the frequency domain of UWB antennas is presented, introducing the important quantities and the parameters for the antenna analysis. This permits to evaluate the influence of the application of filters to the antenna structure. Once the mathematical description is given, the integration of UWB antennas and filters is regarded: firstly, integration of bandpass filters and antennas for selecting a particular frequency interval; secondly, the band-notch antennas are presented, which allow for suppressing narrowband interference, e.g. WLAN devices.

4.1 Performance Criteria Measures of UWB Antennas

In this section the most important parameters to quantify the performance and the quality of UWB antennas are presented, both in the frequency domain and in the time domain.

4.1.1 Frequency Domain

In the frequency domain the antenna's behavior and performance are evaluated based on measurements with continuous wave excitation, which render voltage transfer coefficients that can be interpreted as samples of a transfer function at different frequencies. The parameter that is effectively measured is the transmission coefficient S_{21} between the transmit and the receive antennas.

Let $\mathbf{H}_{Tx}$ be the transfer function of the transmit antenna and $\mathbf{H}_{Rx}$ the transfer function of the receive antenna, in term of the effective antenna height. Let consider a orthonormal polarization basis $\hat{\mathbf{r}}_\eta, \hat{\mathbf{r}}_\zeta$ at the receiver (e.g. the vertical and horizontal polarizations), i.e. $< \hat{\mathbf{r}}_i, \hat{\mathbf{r}}_j >= 1$ for $i = j$ and 0 for $i \neq j$, being i, j either η oder ζ. For U_{Tx} and U_{Rx} being the measured transmitted and received voltages, respectively, the transmission coefficient for a particular polarization η of the receiver, is calculated as (for free space propagation)

$$S_{21}^\eta(\omega, \theta_{Tx}, \psi_{Tx}, \theta_{Rx}, \psi_{Rx}) = U_{Rx}/U_{Tx}$$

$$= \mathbf{H}_{Tx}(\omega, \theta_{Tx}, \psi_{Tx}) \cdot \mathbf{H}_{Rx}^\eta(\omega, \theta_{Rx}, \psi_{Rx}) \cdot \frac{j\omega}{2\pi r c_0} e^{-j\omega r/c_0}$$

$$= H_{Tx}^\eta(\omega, \theta_{Tx}, \psi_{Tx}) \cdot H_{Rx}^\eta(\omega, \theta_{Rx}, \psi_{Rx}) \cdot \frac{j\omega}{2\pi r c_0} e^{-j\omega r/c_0} \qquad (4.1)$$

where $H_{Tx}^\eta =< \mathbf{H}_{Tx}, \hat{\mathbf{r}}_\eta >$, i.e. it is the component of the transmit antenna transfer function for the η polarization of the receiver, $\mathbf{H}_{Rx}^\eta = H_{Rx}^\eta \hat{\mathbf{r}}_\eta$ with H_{Rx}^η the transfer function, in terms of the effective antenna height, of the receive antenna for the η polarization of the receiver, r is the distance between the two antennas and c_0 is the velocity of light. η is the polarization of the receive antenna. In the previous equation the angular dependence of the transfer functions is noted. From the transmit antenna point of view, the receive antenna is positioned in the direction (θ_{Tx}, ψ_{Tx}) in the local coordinate system. Accordingly, for the receive antenna, the transmit antenna is positioned in the direction (θ_{Rx}, ψ_{Rx}) in the local coordinate system.

The receive antenna is assumed to be a reference antenna with two orthogonal polarizations, which in the following, without loss of generality, are assumed to be the horizontal ^{h} and the vertical ^{v} polarizations. In the measurements it does only receive one polarization component which is either the horizontal polarization ^{h} or the vertical polarization ^{v}. The polarization of the transmit antenna is inherently dependent on the angles (θ_{Tx}, ψ_{Tx}). Letting the transmit antenna be the antenna to be investigated (hereafter named AUT, Antenna Under Test) and the receive antenna the reference antenna (hereafter ref) whose parameters are known, it is possible to calculate the transfer function of the AUT, for a particular polarization η (either ^{h} or ^{v}) of the receive reference antenna, from the S_{21} parameter as

$$H_{AUT}^\eta(\omega, \theta_{Tx}, \psi_{Tx}) = \frac{2\pi r c_0}{j\omega} \cdot \frac{S_{21}^\eta(\omega, \theta_{Tx}, \psi_{Tx}, \theta_{Rx}, \psi_{Rx})}{H_{ref}^\eta(\omega, \theta_{Rx}, \psi_{Rx})} \cdot e^{j\omega r/c_0} . \qquad (4.2)$$

The apex η in H_{AUT}^η indicates that the calculated AUT transfer function is the one obtained from the selected η polarization of the receive antenna. In literature, the described method is referred to as two-antenna method for the recovery of the antenna transfer function [6].

The frequency domain parameters for UWB antennas can be calculated starting from the antenna transfer function H. In the following, the most important parameters are summarized. The parameters can be calculated for both polarization components. In the following the polarization apex will be omitted.

Gain It can be directly calculated from the antenna transfer function as

$$G(\omega,\theta,\psi) = \frac{\omega^2}{\pi c_0^2}|H(\omega,\theta,\psi)|^2 \, . \tag{4.3}$$

Starting from this definition, it is possible to obtain the lossless gain defined by the IEEE [60] as

$$G_{\text{IEEE}}(\omega,\theta,\psi) = \frac{G(\omega,\theta,\psi)}{1 - |S_{11}|^2} \tag{4.4}$$

where $|S_{11}|$ is the reflection factor at the antenna input.

Group Delay Time It is defined (according to eq. (2.3) in chapter ?) as

$$\tau_{\text{GDT}}(\omega,\theta,\psi) = -\frac{\partial}{\partial\omega}\angle H(\omega,\theta,\psi) \, . \tag{4.5}$$

Hence, since the antenna transfer function has an angular dependence for each polarization component, also the antenna's radiation parameters, which can be directly calculated from H_{AUT} (such as gain and group delay time) and that permit to characterize the antenna in the frequency domain, are angular dependent.

4.1.1.1 Calculation of the Reference Antenna's Transfer Function

In oder to recover the transfer function of the AUT, the knowledge of the reference antenna's transfer function H_{ref} is necessary.

In order to recover H_{ref}, two identical antennas (hence having the same transfer function), positioned one in front of each other at the distance r, can be used. In this case, eq. (4.1) becomes

$$S_{21} = \frac{e^{(-j\omega r/c_0)}}{2\pi r c_0}j\omega H_{\text{AUT}}^2 \, . \tag{4.6}$$

Hence, it follows

$$H_{\text{AUT}} = \sqrt{\frac{2\pi r c_0}{j\omega}S_{21}e^{(j\omega r/c_0)}} \, . \tag{4.7}$$

With the knowledge of the antennas distance r and the measurement system calibrated to the antenna ports, H_{AUT} can be calculated from an S_{21} measurement by evaluating the complex root with phase unwrapping.

4.1.2 Time Domain

Together with this frequency domain analysis, also a time domain analysis is required to characterize the transient behavior of UWB antennas for pulsed operations.

From the measured transfer function of the AUT, it is possible to derive the AUT transient response $h_{\mathrm{AUT}}(t)$ in the time domain. It is calculated, according to the procedure illustrated in chapter 2, starting from the analytic transfer function defined as

$$H_{\mathrm{AUT}}^{+}(\omega,\theta,\psi) = \begin{cases} 2H_{\mathrm{AUT}}^{+}(\omega,\theta,\psi) & \omega > 0 \\ H_{\mathrm{AUT}}^{+}(\omega,\theta,\psi) & \omega = 0 \\ 0 & \omega < 0 \end{cases} \tag{4.8}$$

and then taking the real part of its inverse Fourier transform $h_{\mathrm{AUT}}^{+}(t,\theta,\psi)$, namely

$$h_{\mathrm{AUT}}(t,\theta,\psi) = \Re\left[h_{\mathrm{AUT}}^{+}(t,\theta,\psi) \right] . \tag{4.9}$$

It has to be noticed that, since the antenna transfer function H_{AUT} depends, for each polarization component, on the angular directions, also the antenna impulse response, h_{AUT}, for each polarization component, depends on the angular directions. Hence, all time domain parameters introduced in chapter 2 (Peak, FWHM, ...) are angular dependent. This means that UWB antennas radiate different pulse shapes in different spatial directions when excited with an impulse, as exemplified in Fig. 4.1, i.e. a spatial distortion on the radiated pulse is introduced.

Due to this angular dependence, it is a crucial problem to investigate the antenna performance with respect to the mutual orientation of the transmit and receive antennas. This problem will be regarded in detail in the next chapter.

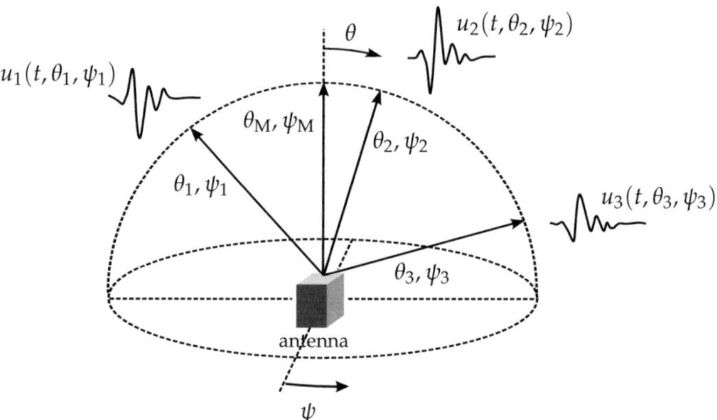

Figure 4.1: Visualization of the effect of the antenna spatial distortion in the time domain, (θ_M, ψ_M) indicates the antenna's main radiation direction.

4.2 Integration of UWB Antennas and Filters

In this section the integration of filters and UWB antennas is investigated. This is important since, as previously introduced, the RF front-end is composed by both the UWB antenna and one or more filters for band selection and/or band rejection. It has to be observed that realizing the antenna and the filter separately, i.e. as two different elements, causes problems of impedance matching at the filter/antenna interface, leakage due to the cable connection of these two elements and also an increment of the overall system dimensions. Integrating the filter into the antenna (either in the antenna ground plane or in the antenna radiating element), helps to decrease the system dimensions. Moreover, it permits to avoid losses due to the antenna/filter connection and leads to better system performance, in particular if the antenna and the filter are designed together, i.e. if co-design is adopted.

In the following different types of integrations are considered, which can be summarized into two main groups: antenna + bandpass filter and antenna + bandstop filter. In the first group a bandpass filter is added to the antenna in order to limit its operational frequency range and hence the bandwidth of the spectrum of the radiated signal. The second group are the so-called Band-Notch UWB antennas, which suppress the radiation of the excitation signal in a specific band, e.g. the 5 GHz WLAN band.

4.2.1 Integration of UWB Antennas and Bandpass Filters

In this section the integration of UWB antennas and bandpass filters is investigated. The filter allows to select a particular frequency interval of the spectrum of the radiated signal. This problematic arises in particular when an antenna has to be used according to a certain regulation (e.g. the EU UWB regulation) and the transmit signal is provided from a pulse generator with a very huge bandwidth, larger than the permitted one.

One technique, which permits to restrict the spectral components to the desired frequency interval, consists in adding a perturbation on the antenna feeding line. This perturbation acts as a bandpass filter.

To exemplify this technique, a particular antenna has been selected, which is suitable to be used according to the FCC regulation. To this antenna, a particular filter structure has been integrated, to permit the antenna to be used in the European context. The developed structure permits to pre-filter the signal in order to select the desired frequency interval $6 - 8.5$ GHz.

The selected antenna is a coplanar antenna [61], fed by a CPW, and it has been optimized for the FCC frequency range. This antenna has been selected because it permits an easier integration of a filter in its radiating element structure and in its ground plane than other typical UWB antennas (such as Vivaldi or Bow-tie).

To the feeding line of this antenna a filter has been introduced, for selecting the EU UWB frequency range. The developed filter has been illustrated in detail in chapter 3 (ref. to section 3.5.5.2). A prototype of the base antenna model and of the integrated system antenna+filter have been etched on the substrate Rogers RO4003, with $\varepsilon_r = 3.38$ and thickness $h = 1.57$ mm (see Fig. 4.2 and Fig. 4.10, left). Measurements have been performed in order to test the behavior of the realized prototypes.

The gain of the antennas has been measured, according to the method illustrated at the beginning of this chapter in section 4.1.1. For these measurements, the antenna was placed in the anechoic chamber with respect to the coordinate system as shown in

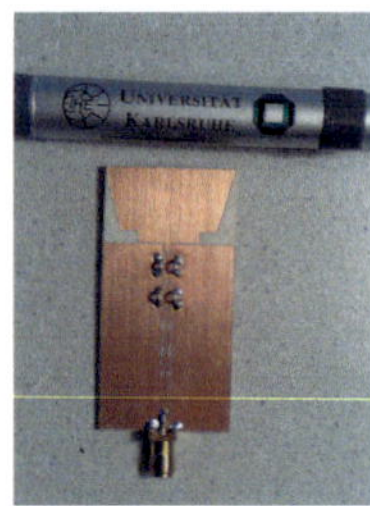

Figure 4.2: The fabricated antenna with the integrated bandpass filter.

Fig. 4.3. For each angular direction ψ, with an accuracy of 4 degrees, the S_{21} parameter between the reference horn antenna and the investigated antenna has been recorded in the frequency interval 0.4 GHz – 20 GHz (801 frequency samples). From these measurements, according to eq. (4.7) the transfer function of the investigated antenna has been recovered. Then, the gain has been calculated as described in 4.1.1.

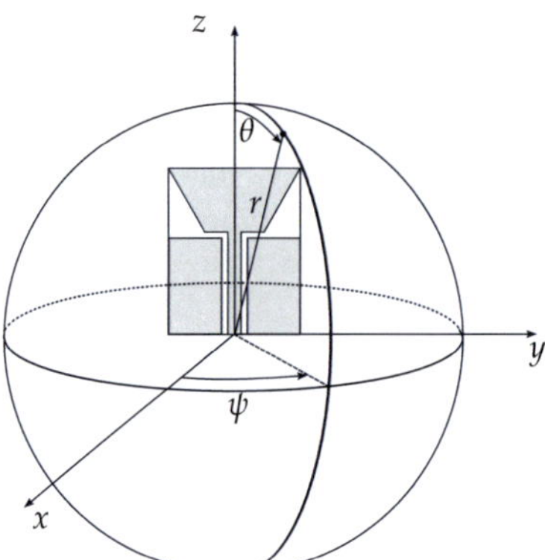

Figure 4.3: Antenna orientation with respect to the coordinate system.

Measurements have been taken for the base antenna and for the antenna with the integrated bandpass filter. The so measured gain in the H-plane, co-polarization component, is plotted in Fig. 4.4 for the base antenna and in Fig. 4.5 for the antenna+bandpass filter. The antenna without filter has high gain in a large frequency interval (approx. 4-10 GHz). Conversely, the antenna with the integrated filter has high gain only in the desired EU UWB frequency range. It has also to be observed that the maximum gain is in this case lower than in the case of the base antenna. This is due to the losses in the filter structure. These losses make the S_{21} parameter of the filter not to be perfectly unitary (0 dB) in the filter passband (ref. to Fig. 3.26).

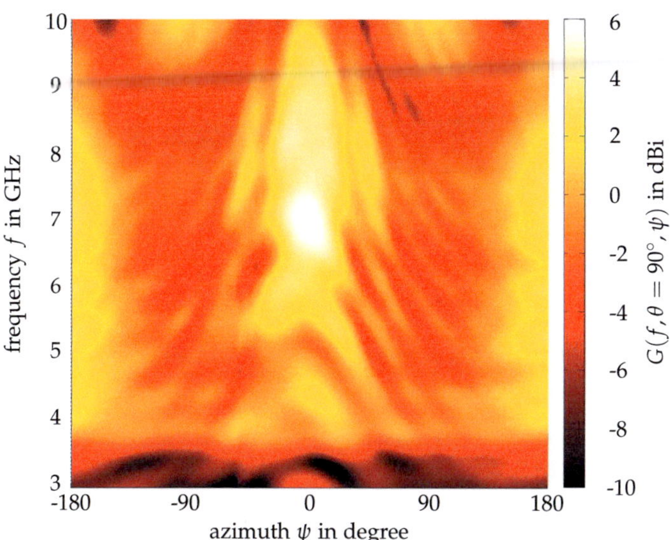

Figure 4.4: Measured gain in the H-plane, co-plarization, for the base antenna.

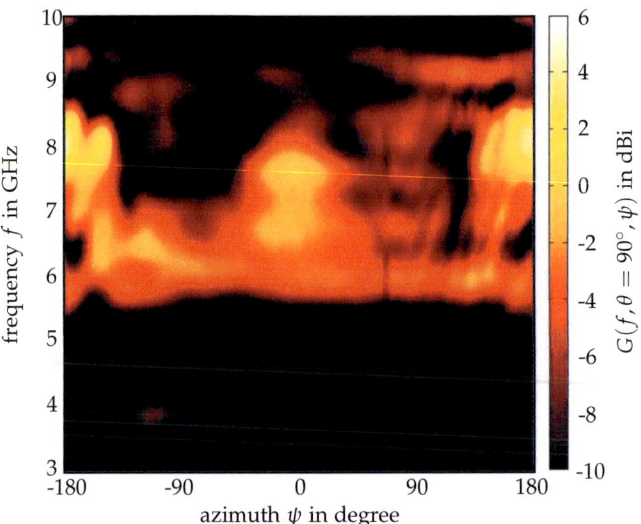

Figure 4.5: Measured gain in the *H*-plane, co-plarization, for the antenna with the integrated bandpass filter.

4.2.2 Integration of UWB Antennas and Bandstop Filters: the Band-Notch UWB Antennas

Band-notch UWB antennas are designed to not operate in a particular frequency interval, and hence the rejection of a certain frequency band is obtained.

The band rejection in the UWB case is necessary for certain applications, in order to reduce the interference between UWB devices and other devices operating in the same band. In fact, in the UWB frequency interval $5 - 6$ GHz there are already existing standards such as IEEE802.11a wireless local area network (WLAN) and HIPERLAN/2. Hence, in order to prevent interference between UWB devices and already existing devices operating in that band, an important goal is the rejection of such already used frequency interval. Consequently, UWB antennas operating in the $3.1 - 10.6$ GHz frequency band with the particularity of presenting a band rejection for the $5 - 6$ GHz interval have been developed, using different methods and techniques to achieve the rejection of the unwanted band.

The aim of this section is to perform a study of different UWB band-notch techniques in order to evaluate their effect and influence on the antenna radiation characteristics (impact on the radiation pattern, on the gain, ...) [37]. Moreover, the time domain behavior of the different UWB band-notch techniques is analyzed, in order to investigate their impact on the transmitted pulse [38].

4.2.2.1 Band-Notch Techniques

It is possible to classify the most commonly used band-notch methods for antenna integration in two main groups. These methods utilize different techniques and approaches to obtain the band rejection. In the following the typical methods are presented and investigated.

Methods acting on the antenna's radiating element The first group comprises all techniques which are based on the addition of a perturbation in the antenna radiating element [36], [62], [63]. Such perturbation usually consists of a slot carved in the antenna radiating element. The effect of such perturbation is to create a destructive interference of the currents at a particular frequency, which can be tuned changing the dimensions and the position of the added perturbation. This technique is exemplified in Fig. 4.6, where a slot carved in the antenna radiating element is the perturbation that permits to achieve the band rejection. In Fig. 4.6, right, the parameters that permit to control the added perturbation are indicated. A particular UWB antenna structure has been selected (ref. to [61]). To this basic antenna structure a slot in the radiated element has been added, as illustrated in Fig. 4.6, right. Through computer simulations it has been possible to assess the effect of the slot on the antenna behavior. It has been seen that the most important parameters for controlling the band rejection are the position and the dimensions of the added slot. The obtained results are summarized in Fig. 4.7, which illustrates the effect of changing the dimensions and the position of the slot (ref. to Fig. 4.6, right) on the input impedance matching of the antenna.

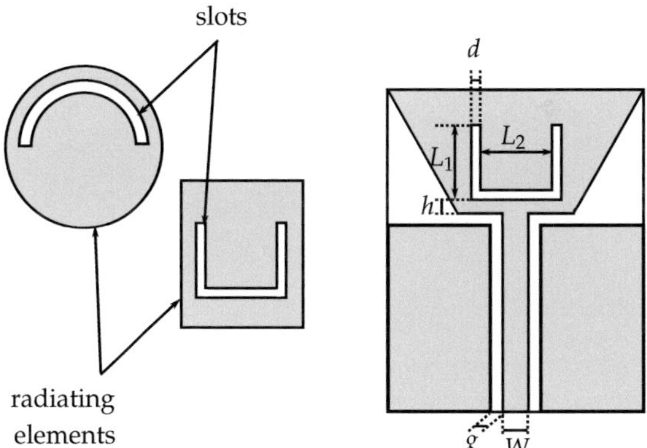

Figure 4.6: Band-notch with methods acting on the antenna's radiating element (left) and the important parameters for controlling the band rejection (right).

Methods acting on the antenna's ground plane or feeding line The second group comprises all techniques, which add perturbations on the antenna feeding line (or on

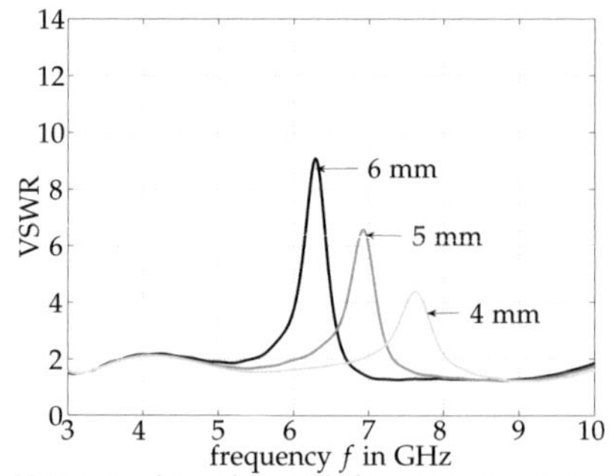

(a) Variation of L_2: the notch frequency increases as L_2 decreases.

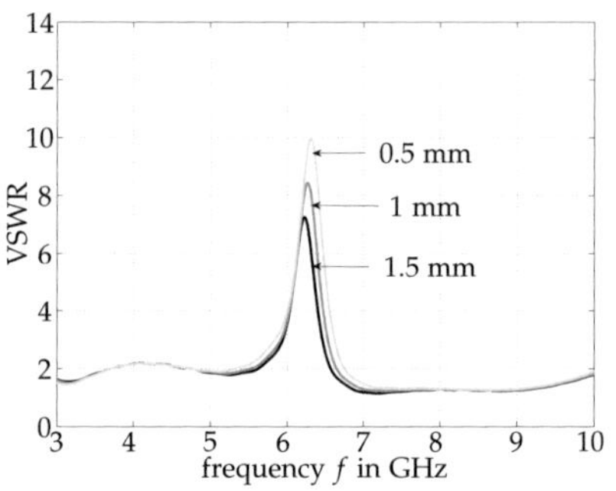

(b) Variation of d: the rejection worsens as d increases.

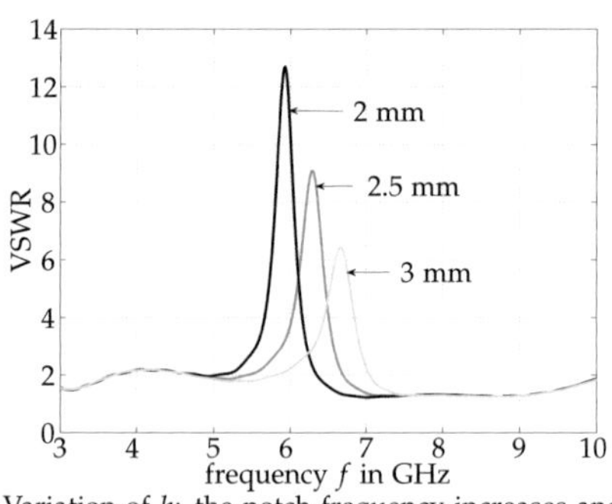

(c) Variation of h: the notch frequency increases and the rejection worsens as h increases.

Figure 4.7: Effect of changing the dimensions of the critical parameters of the slot in the antenna radiating element.

the antenna ground plane) instead of on the antenna radiating element [35], [64]. Such perturbations act as a band stop filter. The aim of such perturbations is to pre-filter the signal in order that the signal that excites the radiating element does not have a spectral component in the unwanted frequency band. This technique is exemplified in Fig. 4.8, where, on the right side, the most critical dimensions for controlling the band rejection are indicated. In order to assess the effect of this method, a particular UWB antenna structure (the same as in the previous case, in order to allow for comparison) has been selected. To this structure slots in the antenna ground plane, acting as a bandstop filter, have been inserted. The obtained structure is shown in Fig. 4.8, right. In order to implement the required bandstop filter, the previously introduced filter design (ref. to section 3.5.1) has been applied. With computer simulations it has been possible to assess the effect of changing the position and dimensions of the added slots. The obtained results are illustrated in Fig. 4.9, where the effect of the different parameters on the band rejection quality and on the filter tuning is shown.

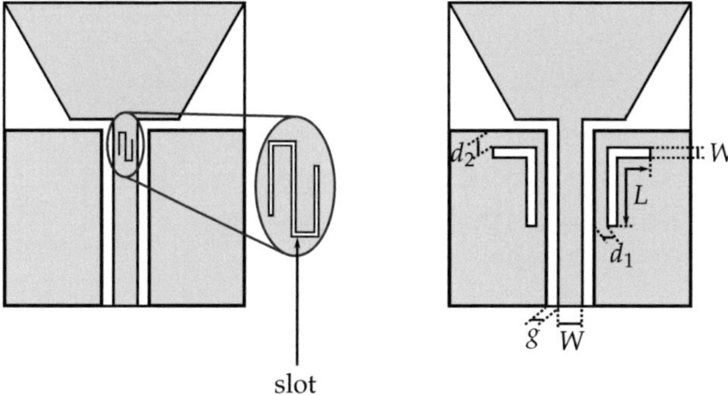

Figure 4.8: Band-notch filter integration with methods acting on the antenna's feeding line (left) and on the antenna ground plane together with the important dimensions for controlling the band rejection (right).

In order to investigate the characteristics and the effects in the frequency domain and in the time domain of these two different methods, the particular antenna topology, regarded in the computer simulations, has been fabricated. Starting from this basic antenna structure (hereafter called antenna *A*), two different antennas have been fabricated, applying to one antenna the perturbation in the radiating element (hereafter called antenna *B*) and to the other one the pre-filtering structure (antenna *C*). The three fabricated antennas are shown in Fig. 4.10. In both cases, the dimensions and position of the perturbations have been optimized with computer simulations, starting from the results previously shown, in order to obtain the desired rejected frequency and the best achievable rejection capability with the selected method. The aim of the filter design has been the rejection of the 5.8 GHz WLAN band.

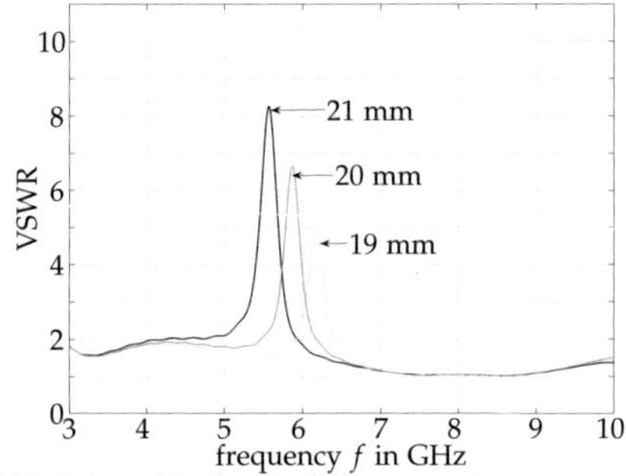

(a) Variation of L: the notch frequency increases and the rejection decreases as L decreases.

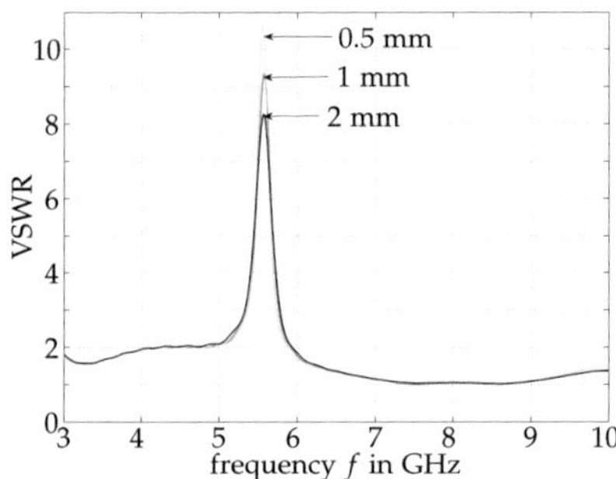

(b) Variation of d_1: the rejection worsens as d_1 increases.

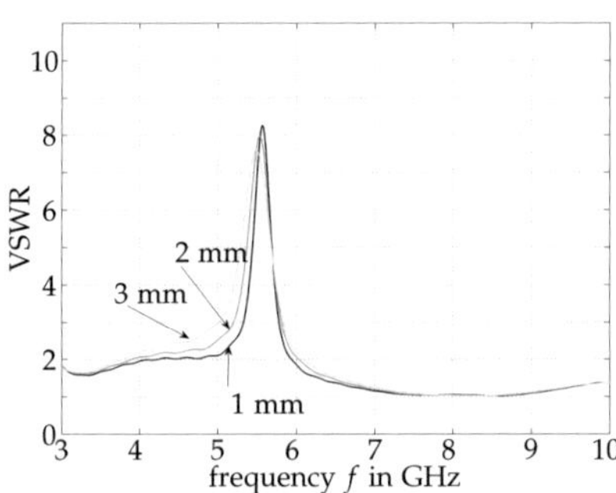

(c) Variation of d_2: the rejection worsens as d_2 increases.

Figure 4.9: Effect of changing the dimensions of the critical parameters of the slot in the antenna ground plane.

Figure 4.10: The three fabricated antennas: base antenna (left), antenna with band-notch through perturbation in the radiating element (center), antenna with band-notch through perturbation in the ground plane (right).

4.2.2.2 Frequency Domain Analysis

In Fig. 4.11 the measured gain for both the notch antenna configurations (H-plane, co-polarization) is plotted, while the measured gain for the base antennas has been shown in Fig. 4.4. The antenna configuration C has better performance with respect to the antenna B, since the notch is deeper. Furthermore, the gain around the notched area is similar compared to the base antenna A, i.e. introducing the slot in the ground plane slightly affects the gain except at the notch frequency. On the other hand, in the antenna configuration B the gain around the notched area is highly affected by the slot. This can be even better observed in Fig. 4.13-4.15 where the gain plots for the main beam direction ($\theta, \psi = 0°$), co-polarization, are illustrated.

This different behavior between the antennas B and C can be also observed in the gain pattern, which is plotted in Fig. 4.16 for all antennas (H-plane, co-polarization) at the frequency $f = 8$ GHz. At this frequency the gain patterns of all three antennas are approximately identical, hence the antenna configurations B and C do not have a deep influence at a frequency far away from the notch. On the other hand, both antennas B and C present a very low gain pattern at the notch frequency of 5.8 GHz (see Fig. 4.17). However, antenna B has two small lobes in the main radiation direction ($0°$ and $180°$), which are not present in the case of antenna C.

This different behavior can be explained by examining the current distribution on the two antennas in different cases: at the notch frequency and at a frequency far away from the notch ($f = 8$ GHz). For $f = 8$ GHz, both slot configurations slightly affect the distribution of the currents. However, if the antenna is excited with the resonance frequency of the slot, at the notch frequency, in the case of antenna B the currents resonate on the radiating element and this produces the small lobes in the antenna main radiation direction. On the other hand, in the case of antenna C, the slots in the ground plane act as a filter which blocks the currents before they arrive on the radiating element, which is consequently not excited. Hence here do not occur any sidelobe [37].

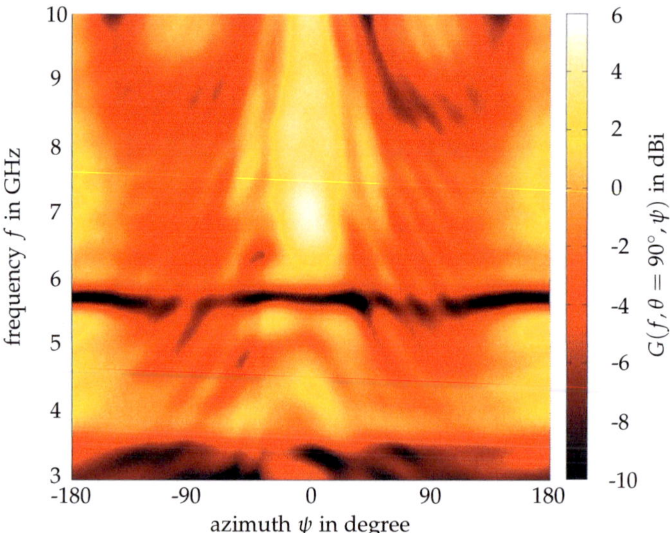

Figure 4.11: Measured gain in the H-plane, co-plarization, for the antenna with slot in the radiating element.

4.2.2.3 Time Domain Analysis

The antenna impulse response for both band notch antenna configurations has been measured, together with the impulse response of the base antenna for comparison.

In Fig. 4.18-4.20 the measured impulse responses of the three antennas in the H-plane (co-polarized component) are plotted against the time and the azimuth angle. Comparing Fig. 4.19 and 4.20 with Fig. 4.18, it can be recognized that the addition of a slot in the antenna radiating element or in the antenna ground plane influences the time domain behavior of the antenna. By integrating the filter, the ringing increases and the peak value decreases. In order to better quantify this influence, the impulse responses of the antennas in the main beam direction are analyzed.

The measured transient responses in the main beam direction (co-polarization) of the three fabricated antennas are shown in Fig. 4.21-4.23. They correspond to vertical cuts of the plots in Fig. 4.18- 4.20 for $\psi = 0°$. The two notch techniques show different effects in the time domain. The addition of a slot in the antenna radiating element (Fig. 4.22) causes the peak of the antenna impulse response (in the main beam direction) to significantly decrease ($P = 0.1548$ m/ns) with respect to the base antenna ($P = 0.1705$ m/ns, Fig. 4.21). Moreover, there is a small increment of the ringing (for $r = 10\%$, $\tau_{0.1} = 0.924$ ns) with respect to the base antenna ($\tau_{0.1} = 0.850$ ns). On the other hand (Fig. 4.23), filtering the signal with a filter in the antenna ground plane,

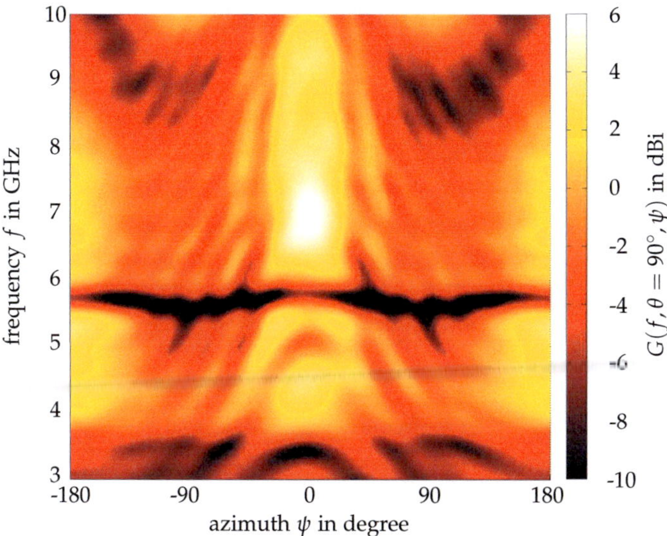

Figure 4.12: Measured gain in the *H*-plane, co-plarization, for the antenna with slot in the ground plane.

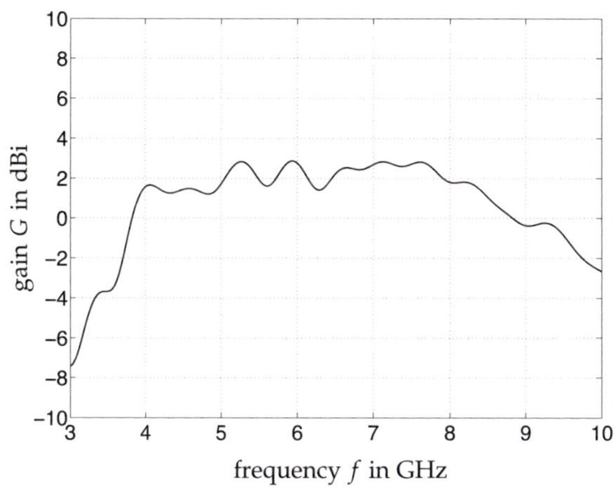

Figure 4.13: Measured gain in the main beam direction, co-plarization, for the base antenna.

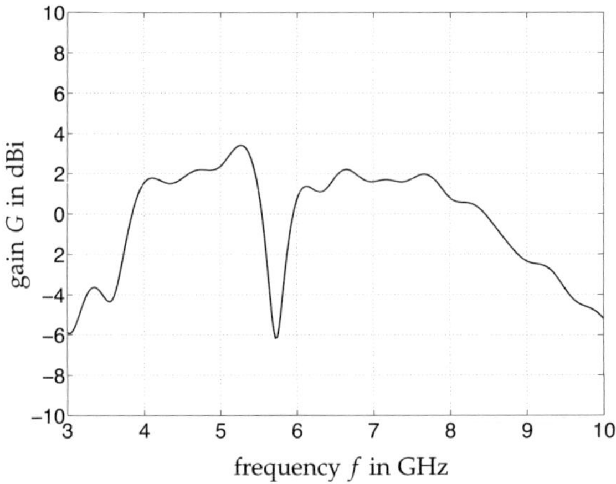

Figure 4.14: Measured gain in the main beam direction, co-plarization, for the antenna with slot in the radiating element.

causes the ringing duration to slightly increase ($\tau_{0.1} = 1.029$ ns) with respect to the base antenna, but the peak value is clearly higher ($P = 0.17$ m/ns) compared to antenna B and nearly identical to the result for antenna A without filter. The measured time domain parameters for the three antennas in the main beam direction are summarized in Tab. 4.1 for comparison.

Table 4.1: Measured Time Domain Parameters (Main Beam Direction)

Antenna	Peak P	τ_{FWHM}	$\tau_{0.1}$
A: Without filter	0.1705 m/ns	105 ps	0.850 ns
B : Slot in the radiating element	0.1584 m/ns	142 ps	0.924 ns
C : Slots in the ground plane	0.1700 m/ns	110 ps	1.029 ns

Consequently, the application of a perturbation in the antenna radiating element provokes a relatively high decrement of the peak (-8% with respect to the base antenna case) and a high increment of the FWHM ($+35\%$ with respect to the base antenna). On the other hand, the application of the slots in the antenna ground plane, i.e. pre-filtering the signal, causes a high increment of the ringing ($+21\%$ with respect to the base antenna case), while the variation of the peak (-0.5%) and of the FWHM ($+4\%$) compared to the base antenna case is negligible. Hence, regarding all time domain parameters, the integration of the slots into the antenna ground plane results in a clearly better time domain performance compared to the integration in the antenna radiating element.

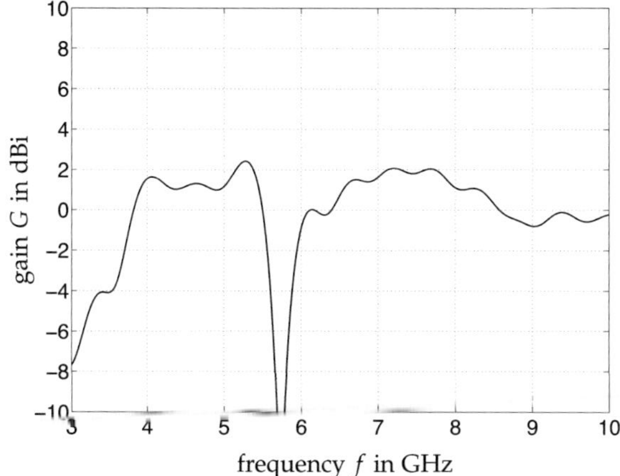

Figure 4.15: Measured gain in the main beam direction, co-plarization, for the antenna with slots in the ground plane.

Hence, applying a perturbation on the antenna radiating element has a deep impact on the time domain behavior of the antenna, since the peak highly decreases and the FWHM highly increases with respect to the base antenna. The case with the perturbation added in the antenna ground plane shows, except for the slight increase of the ringing duration, only negligible deterioration of the antenna performance and hence should be the preferred realization for practical antenna designs with integrated band notch filtering.

4.3 Conclusion

In this chapter the effect of the integration of filters with UWB antennas has been investigated. Two different kinds of integration have been considered: integration of antenna and bandpass filter for frequency selection and integration of antenna and bandstop filter for frequency suppression. It has been seen that these filters can be applied either in the antenna ground plane, hence acting on the antenna feeding line, or in the antenna radiating element. Through computer simulations it has been possible to assess the influence of the added perturbations (dimensions and position), which act as filters, either in the antenna radiating element or in the antenna ground plane/feeding line, on the band selection of the realized structure. Prototypes with filters in antenna feeding line, in the ground plane and in the radiating element have been fabricated and tested. From measurement results of the base antenna structure (i.e. without perturbations) and of the prototypes with perturbations, the frequency and time domain behaviors of the different antenna+filter structures and the base an-

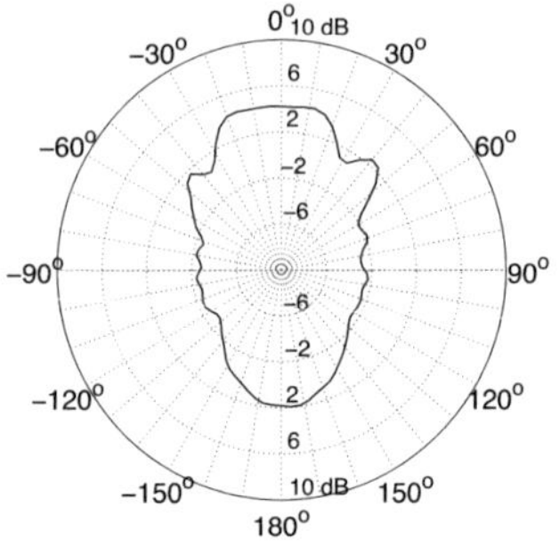

(a) Base antenna

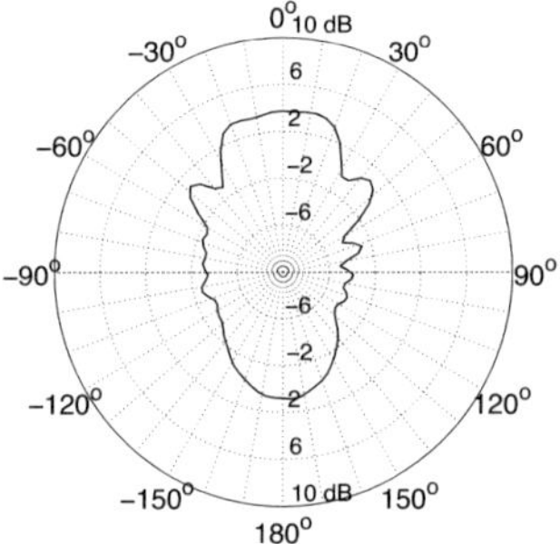

(b) Antenna with slot in the radiating element

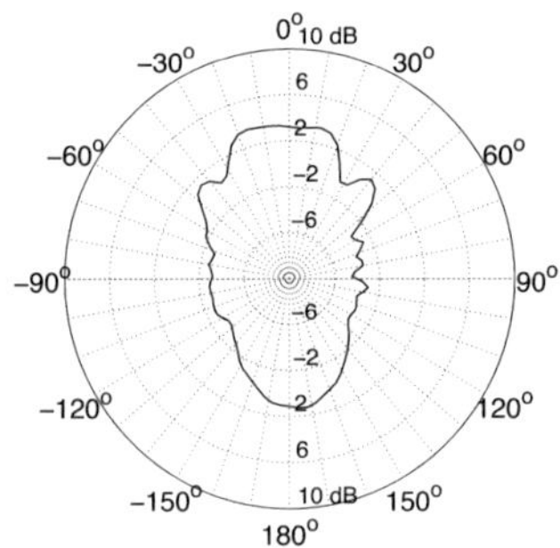

(c) Antenna with slots in the ground plane

Figure 4.16: Measured gain patterns of the three antennas at 8 GHz, H-plane, co-polarization.

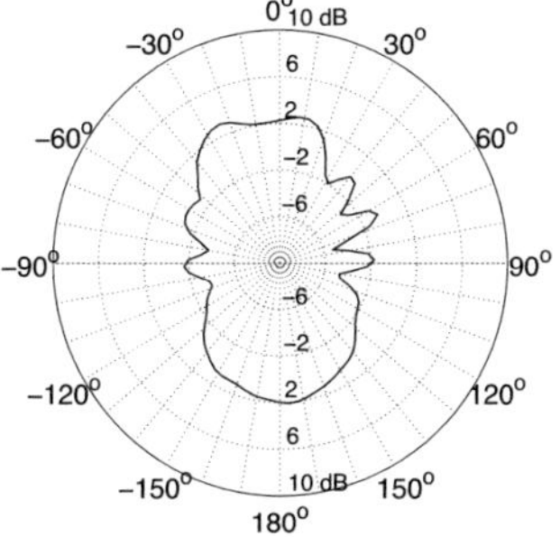

(a) Base antenna

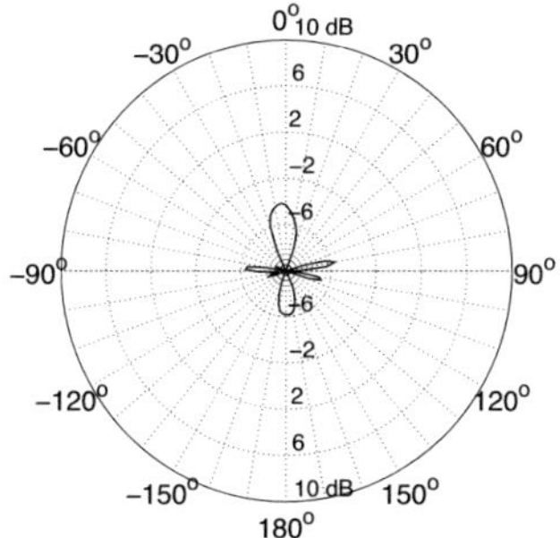

(b) Antenna with slot in the radiating element

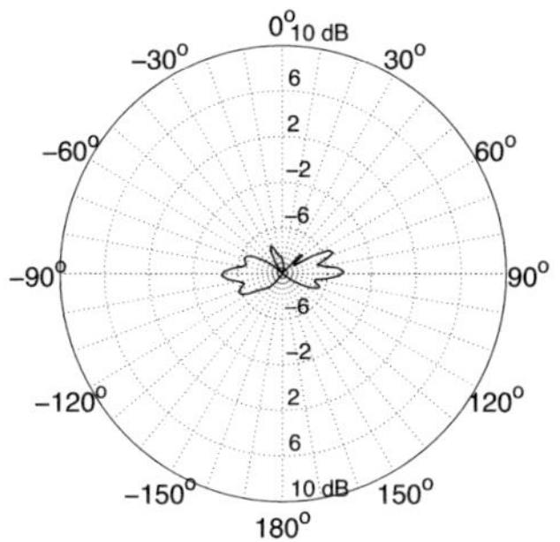

(c) Antenna with slots in the ground plane

Figure 4.17: Measured gain patterns of the three antennas at 5.8 GHz, H-plane, co-polarization.

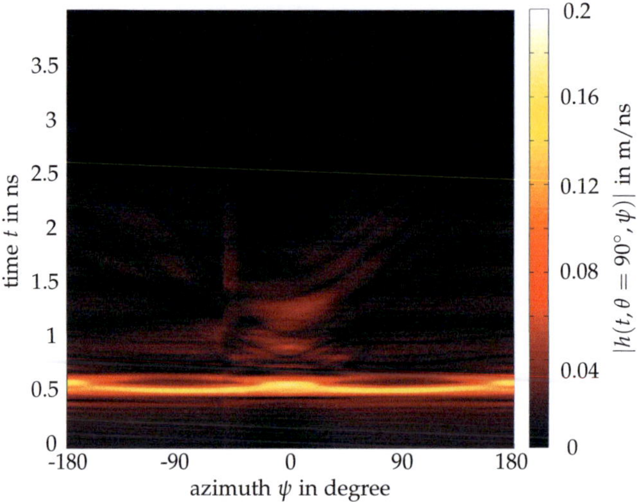

Figure 4.18: Impulse response envelope of the base antenna, H-plane, co-polarization.

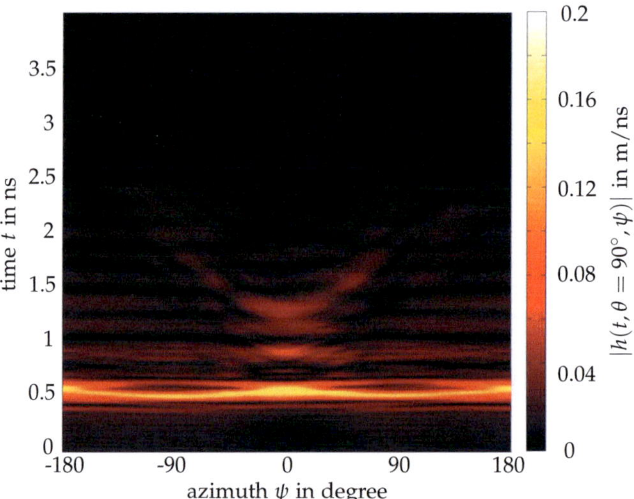

Figure 4.19: Impulse response envelope of the antenna with slot in the radiating element, H-plane, co-polarization.

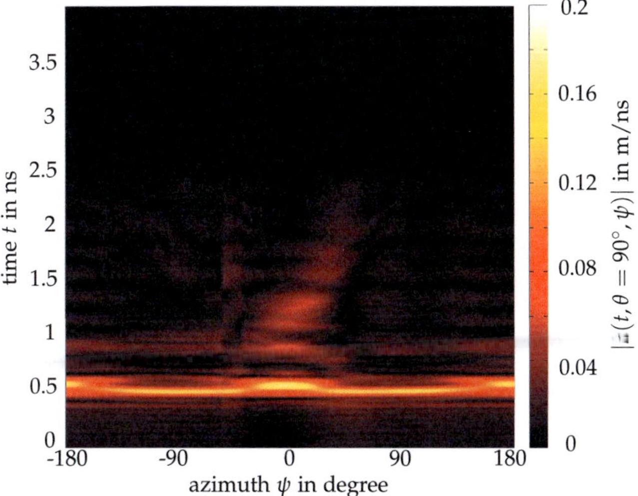

Figure 4.20: Impulse response envelope of the antenna with slots in the ground plane, *H*-plane, co-polarization.

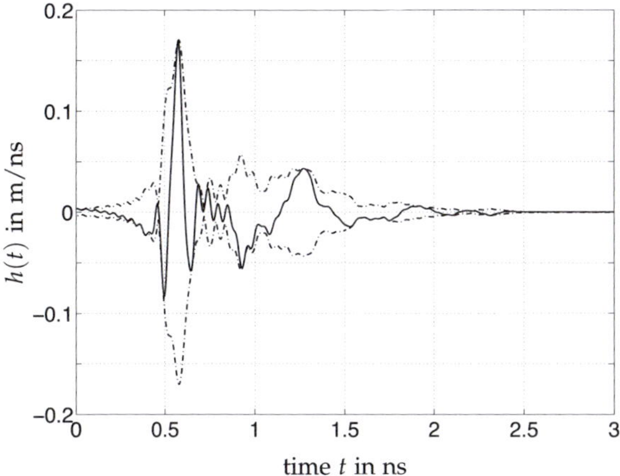

Figure 4.21: Measured impulse response of the base antenna (solid line) and its envelope (dotted line), main beam direction, co-polarization.

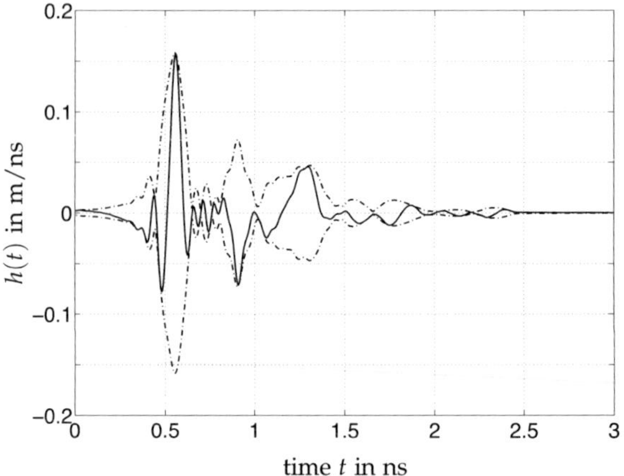

Figure 4.22: Measured impulse response of the antenna with slot in the radiating element (solid line) and its envelope (dotted line), main beam direction, co-polarization.

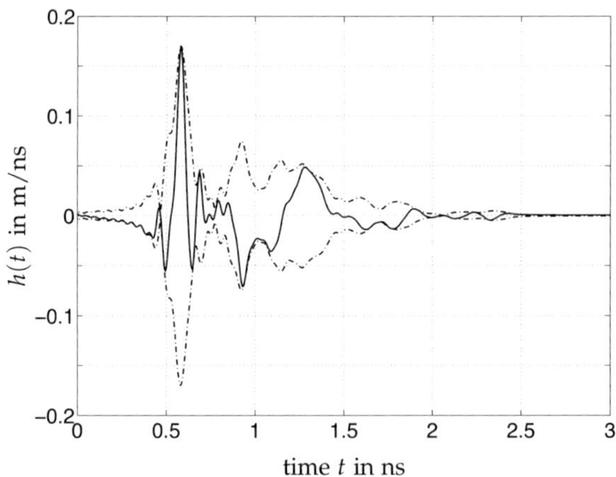

Figure 4.23: Measured impulse response of the antenna with slots in the ground plane (solid line) and its envelope (dotted line), main beam direction, co-polarization.

tenna have been compared. It has been derived that methods acting on the antenna's feeding line have better time domain and frequency domain performance with respect to methods acting on the antenna's radiating element.

5 UWB Radio Link

In this chapter the analytical description of a complete UWB radio link is regarded, both in the frequency domain and in the time domain. The aim is to introduce novel quality criteria, which allow for the assessment of the signal distortion caused by the involved components. In this context the behavior of the antennas is of particular interest, since in addition to their frequency dependent behavior they also act as a spatial filter and show different properties in the various angular directions.

From the receiver point of view, the condition for obtaining the optimum signal to noise ratio is investigated. This leads to an analysis in the time domain of the distortion that the transmitted signal undergoes in the different radiation directions with respect to the signal transmitted in the main beam direction. This permits to investigate the pulse preserving capability in the different radiation directions of UWB antennas. For this purpose an analysis of the correlation properties of the transmitted pulse shapes is performed in the whole space using a spherical representation. A quality criterion based on fidelity is derived to identify and quantify the areas where the antenna has a good time domain behavior. By these investigations it is possible to determine the size of the spatial regions where the transmitted signal satisfies given fidelity properties. Moreover, a joint criterion for the assessment of both the signal distortion and the radiated peak power will be developed.

The chapter is organized as follows. Firstly, the whole UWB link is described both in the time domain and in the frequency domain. Then, the receiver is regarded in order to find criteria that permit to assess the quality of the UWB link from the perspective of the receiver, taking into account the non-idealities of the antennas. Finally, two different quality criteria will be defined. The introduced criteria permit to quantify the worsening of the signal to noise ratio of an UWB radio link due to the non-ideal behavior of the transmit and receive antennas. Moreover, they allow also for the individuation of the spatial regions where an antenna radiates signals, which are lowly distorted with respect to the signal radiated in the main beam direction.

5.1 The UWB Radio Link Components

In Fig. 5.1 a simplified UWB radio link is shown. At the transmitter side it is composed by a pulse generator and the transmit antenna. At the receiver side by the receive antenna and a sampling device that digitalizes the received signal after an analog preprocessing has been performed. In between there is the UWB channel. In the following, a line of sight link is assumed.

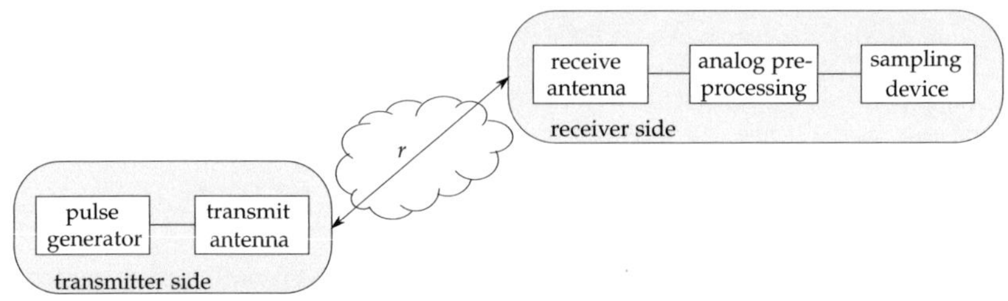

Figure 5.1: UWB radio link block scheme.

5.1.1 Time Domain Description

In the following the time domain description of the UWB radio link is regarded. Let $u_{\text{Tx}}(t)$ be the pulse from the pulse generator and $\mathbf{h}_{\text{Tx}}(t, \theta_{\text{Tx}}, \psi_{\text{Tx}}) = h_{\text{Tx}}^{\theta}(t, \theta_{\text{Tx}}, \psi_{\text{Tx}})\hat{\mathbf{r}}_{\theta_{\text{Tx}}} + h_{\text{Tx}}^{\psi}(t, \theta_{\text{Tx}}, \psi_{\text{Tx}})\hat{\mathbf{r}}_{\psi_{\text{Tx}}}$ be the impulse response of the transmit antenna, which is angular dependent. $h_{\text{Tx}}^{\theta}(t, \theta_{\text{Tx}}, \psi_{\text{Tx}})$ is the component in the direction of the unit vector $\hat{\mathbf{r}}_{\theta_{\text{Tx}}}$ and $h_{\text{Tx}}^{\psi}(t, \theta_{\text{Tx}}, \psi_{\text{Tx}})$ is the component in the direction of the unit vector $\hat{\mathbf{r}}_{\psi_{\text{Tx}}}$ of the antenna impulse response, according to the coordinate system shown in Fig. 5.2. Since the transmitted wave is a TEM wave, the component in the radial direction is not present. The relationship between the radiated electrical filed $\mathbf{e}_{\text{Tx}}$ at a certain distance r from the transmitter and in the generic angular direction $(\theta_{\text{Tx}}, \psi_{\text{Tx}})$ is then modeled as (in the far-field region)

$$\frac{\mathbf{e}_{\text{Tx}}(t, \theta_{\text{Tx}}, \psi_{\text{Tx}}, r)}{\sqrt{Z_0}} = \frac{1}{r}\delta\left(t - \frac{r}{c_0}\right) * \frac{1}{2\pi c_0}\mathbf{h}_{\text{Tx}}(t, \theta_{\text{Tx}}, \psi_{\text{Tx}}) * \frac{\partial}{\partial t}\frac{u_{\text{Tx}}(t)}{\sqrt{Z_{\text{Tx}}}} \qquad (5.1)$$

where $Z_0 = 120\pi\Omega$ is the free-space impedance, Z_{Tx} is the reference impedance at the antenna connector (assumed frequency independent) and the derivation is caused by the transmit antenna, resulting from the reciprocity theorem, as clarified in [46]. Moreover, the term $(\theta_{\text{Tx}}, \psi_{\text{Tx}})$ indicates the angular dependence of the antenna behavior, as discussed in the previous chapter. The term

$$h_{\text{Ch}}(t, r) = \frac{1}{r}\delta\left(t - \frac{r}{c_0}\right) \qquad (5.2)$$

represents the channel impulse response.

Let the receive antenna, with impulse response

$$\mathbf{h}_{\text{Rx}} = h_{\text{Rx}}^{\theta}(t, \theta_{\text{Rx}}, \psi_{\text{Rx}})\hat{\mathbf{r}}_{\theta_{\text{Rx}}} + h_{\text{Rx}}^{\psi}(t, \theta_{\text{Rx}}, \psi_{\text{Rx}})\hat{\mathbf{r}}_{\psi_{\text{Rx}}}$$

be positioned at a distance r_0 from the transmitter and at the angular position $(\theta_{\text{Rx}}, \psi_{\text{Rx}})$ in the angular coordinate system of the receiver, as illustrated in Fig. 5.2. $h_{\text{Rx}}^{\theta}(t, \theta_{\text{Rx}}, \psi_{\text{Rx}})$ is the component in the direction of the unit vector $\hat{\mathbf{r}}_{\theta_{\text{Rx}}}$ and $h_{\text{Rx}}^{\psi}(t, \theta_{\text{Rx}}, \psi_{\text{Rx}})$ is the component in the direction of the unit vector $\hat{\mathbf{r}}_{\psi_{\text{Rx}}}$ of the antenna impulse response, basing on the coordinate system shown in Fig. 5.2. According to the vectors orientation

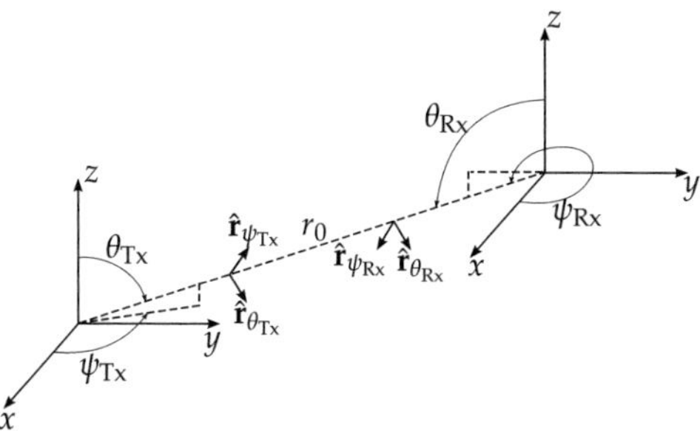

Figure 5.2: Coordinates of the UWB radio link and unit vectors $\hat{\mathbf{r}}_{i_{\text{Tx}}}$ at the transmitter and $\hat{\mathbf{r}}_{i_{\text{Rx}}}$ at the receiver coordinate system, $i = \theta, \psi$.

in Fig. 5.2, the following relationships[1] between the unit vectors at the transmitter side and at the receiver side are valid:

$$\begin{aligned}
< \hat{\mathbf{r}}_{\theta_{\text{Rx}}}, \hat{\mathbf{r}}_{\theta_{\text{Tx}}} > &= 1 \\
< \hat{\mathbf{r}}_{\psi_{\text{Rx}}}, \hat{\mathbf{r}}_{\psi_{\text{Tx}}} > &= -1 \\
< \hat{\mathbf{r}}_{\theta_i}, \hat{\mathbf{r}}_{\psi_j} > &= 0 \quad \text{with } i, j = \text{Tx or Rx}
\end{aligned} \tag{5.3}$$

As described in [8], the received signal $u_{\text{Rx}}(t)$ is given by

$$\begin{aligned}
\frac{u_{\text{Rx}}(t)}{\sqrt{Z_{\text{Rx}}}} &= \mathbf{h}_{\text{Rx}}(t, \theta_{\text{Rx}}, \psi_{\text{Rx}}) * \mathbf{e}_{\text{Tx}}(t, \theta_{\text{Tx}}, \psi_{\text{Tx}}, r_0) \\
&= \mathbf{h}_{\text{Rx}}(t, \theta_{\text{Rx}}, \psi_{\text{Rx}}) * \frac{1}{r_0} \delta\left(t - \frac{r_0}{c_0}\right) * \frac{1}{2\pi c_0} \mathbf{h}_{\text{Tx}}(t, \theta_{\text{Tx}}, \psi_{\text{Tx}}) * \frac{\partial}{\partial t} \frac{u_{\text{Tx}}(t)}{\sqrt{Z_{\text{Tx}}}}
\end{aligned} \tag{5.4}$$

where Z_{Rx} is the reference impedance at the receive antenna connector (assumed frequency independent). For the transmit antenna the direction of the receive antenna is denoted by $(\theta_{\text{Tx}}, \psi_{\text{Tx}})$ in the local coordinate system. Accordingly, for the receive antenna the direction of the transmit antenna is denoted by $(\theta_{\text{Rx}}, \psi_{\text{Rx}})$, i.e., the transmit signal is impinging on the receive antenna from the direction $(\theta_{\text{Rx}}, \psi_{\text{Rx}})$. Basing on the previously stated relationships (5.3) between the unit vectors, the convolution between the impulse response of the receive antenna $\mathbf{h}_{\text{Rx}}$ and the transmit antenna $\mathbf{h}_{\text{Tx}}$ has to be interpreted in the following way

$$\mathbf{h}_{\text{Rx}} * \mathbf{h}_{\text{Tx}} = \left(h_{\text{Rx}}^{\theta} * h_{\text{Tx}}^{\theta}\right) < \hat{\mathbf{r}}_{\theta_{\text{Rx}}}, \hat{\mathbf{r}}_{\theta_{\text{Tx}}} > + \left(h_{\text{Rx}}^{\psi} * h_{\text{Tx}}^{\psi}\right) < \hat{\mathbf{r}}_{\psi_{\text{Rx}}}, \hat{\mathbf{r}}_{\psi_{\text{Tx}}} > . \tag{5.5}$$

5.1.2 Frequency Domain Description

In the frequency domain eq. (5.1) can be rewritten as

$$\frac{\mathbf{E}_{\text{Tx}}(f, \theta_{\text{Tx}}, \psi_{\text{Tx}}, r)}{\sqrt{Z_0}} = \frac{1}{r} e^{-j2\pi f r/c_0} \cdot \mathbf{H}_{\text{Tx}}(f, \theta_{\text{Tx}}, \psi_{\text{Tx}}) \cdot \frac{1}{2\pi c_0} j\omega \frac{U_{\text{Tx}}(f)}{\sqrt{Z_{\text{Tx}}}} \tag{5.6}$$

---

[1]Given two vector $\mathbf{r}_1$ and $\mathbf{r}_2$, the scalar product between them is defined as $< \mathbf{r}_1, \mathbf{r}_2 > = |\mathbf{r}_1| \cdot |\mathbf{r}_1| \cos(\alpha)$, where α is the angle comprised between the direction of the two vectors.

where $\mathbf{H}_{\text{Tx}}(f,\theta_{\text{Tx}},\psi_{\text{Tx}}) = H_{\text{Tx}}^{\theta}(f,\theta_{\text{Tx}},\psi_{\text{Tx}})\hat{\mathbf{r}}_{\theta_{\text{Tx}}} + H_{\text{Tx}}^{\psi}(f,\theta_{\text{Tx}},\psi_{\text{Tx}})\hat{\mathbf{r}}_{\psi_{\text{Tx}}}$ is the antenna transfer function in the generic angular direction $(\theta_{\text{Tx}},\psi_{\text{Tx}})$, with $H_{\text{Tx}}^{\theta}(f,\theta_{\text{Tx}},\psi_{\text{Tx}})$ the component in the direction of the unit vector $\hat{\mathbf{r}}_{\theta_{\text{Tx}}}$ and $H_{\text{Tx}}^{\psi}(f,\theta_{\text{Tx}},\psi_{\text{Tx}})$ the component in the direction of the unit vector $\hat{\mathbf{r}}_{\psi_{\text{Tx}}}$. $U_{\text{Tx}}(f)$ is the spectrum of the generated signal $u_{\text{Tx}}(t)$ and the term $j\omega = j2\pi f$ is due to the differentiation in the time domain.

Similarly, the received signal at a distance r_0 is given by [46]

$$\frac{U_{\text{Rx}}(f)}{\sqrt{Z_{\text{Rx}}}} = \mathbf{H}_{\text{Rx}}(f,\theta_{\text{Rx}},\psi_{\text{Rx}}) \cdot \frac{1}{r_0}e^{-j2\pi f r_0/c_0} \cdot \mathbf{H}_{\text{Tx}}(f,\theta_{\text{Tx}},\psi_{\text{Tx}}) \cdot \frac{1}{2\pi c_0}j\omega\frac{U_{\text{Tx}}(f)}{\sqrt{Z_{\text{Tx}}}} \qquad (5.7)$$

where $\mathbf{H}_{\text{Rx}}(f,\theta_{\text{Rx}},\psi_{\text{Rx}}) = H_{\text{Rx}}^{\theta}(f,\theta_{\text{Rx}},\psi_{\text{Rx}})\hat{\mathbf{r}}_{\theta_{\text{Rx}}} + H_{\text{Rx}}^{\psi}(f,\theta_{\text{Rx}},\psi_{\text{Rx}})\hat{\mathbf{r}}_{\psi_{\text{Rx}}}$ is the transfer function of the receive antenna. The multiplication in eq. (5.7) has to be interpreted as (ref. to eq. (5.3) and (5.5))

$$\mathbf{H}_{\text{Rx}} * \mathbf{H}_{\text{Tx}} = \left(H_{\text{Rx}}^{\theta} \cdot H_{\text{Tx}}^{\theta}\right) < \hat{\mathbf{r}}_{\theta_{\text{Rx}}}, \hat{\mathbf{r}}_{\theta_{\text{Tx}}} > + \left(H_{\text{Rx}}^{\psi} \cdot H_{\text{Tx}}^{\psi}\right) < \hat{\mathbf{r}}_{\psi_{\text{Rx}}}, \hat{\mathbf{r}}_{\psi_{\text{Tx}}} > . \qquad (5.8)$$

5.1.3 The Impulse Radio Receiver

In this section the signal processing at the receiver side is analyzed. Firstly, the optimal case is taken into consideration. Then, the practical case is investigated quantifying the worsening of the system with respect to the optimal case.

5.1.3.1 The Matched Filter Receiver

The aim of the matched filter receiver is to maximize the instantaneous power of the signal component at the receiver compared to the average noise[2] power at a particular time t_0, i.e. it maximizes the signal to noise ration (SNR) at the receiver in presence of additive white noise [45]. This kind of receiver is implemented through a matched filter with impulse response $h_{\text{MF}}(t)$. The received signal $u_{\text{Rx}}(t)$ is convolved with the matched filter impulse response obtaining

$$y(t) = u_{\text{Rx}}(t) * h_{\text{MF}}(t) \qquad (5.9)$$

and a decision is taken about the transmitted symbol based on the value of the convolution output $y(t_0)$, i.e. sampled at the time $t = t_0$, as shown in Fig. 5.3.

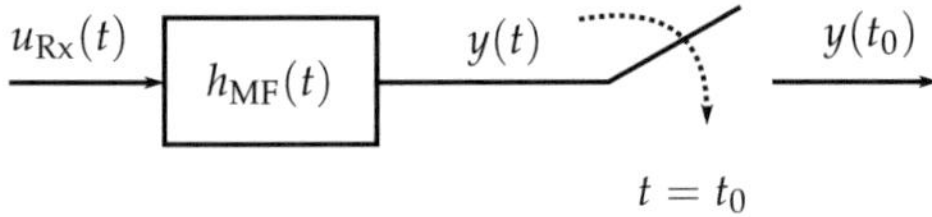

Figure 5.3: Block scheme of a matched filter receiver.

---

[2]The noise is assumed additive white gaussian (AWGN).

According to signal theory [66], the matched filter that permits to maximize the SNR at the time t_0 is

$$h_{\mathrm{MF}}(t) = k \cdot u_{\mathrm{Rx}}^*(t_0 - t) \tag{5.10}$$

where the apex * indicates the conjugate complex and k is a multiplicative constant. In the frequency domain, the previous equation corresponds to[3]

$$H_{\mathrm{MF}}(t) = k \cdot U_{\mathrm{Rx}}^*(f) e^{-j2\pi f t_0} . \tag{5.11}$$

Hence, eq. (5.10) and (5.11) give the optimum filter which permits to maximize the SNR at the receiver.

5.1.3.2 The Correlation Receiver

In practical realizations it is difficult to construct a suitable matched filter, since it is difficult to design an analog circuit that exactly delivers the required impulse response. Furthermore, in order to maximize the SNR, the filter impulse response has to be always matched to the received signal, which is direction-dependent, but a filter realized as analog circuit can only posses one fixed impulse response. Therefore, the receiver implementation through the correlation of the received signal with a reference pulse template is preferred. The template is usually chosen as the expected shape of the received pulse, i.e. it is the idealized received signal in the case of a delay channel (with impulse response $h_{\mathrm{Ch}}^{\mathrm{id}}(t)$, ref. to eq. (5.2)) and with the antennas aligned to their main beam direction. Consequently, under these hypotheses, the template function of the receiver results

$$h_{\mathrm{ref}}(t) = \mathbf{h}_{\mathrm{Rx}}^{\mathrm{M}}(t) * h_{\mathrm{Ch}}^{\mathrm{id}}(t) * \mathbf{h}_{\mathrm{Tx}}^{\mathrm{M}}(t) * \frac{\partial}{\partial t} u_{\mathrm{Tx}}(t) \tag{5.12}$$

where the apex ^{M} indicates the main beam direction of each antenna. It has to be pointed out that the assumed template is coincident to a matched filter, i.e. $h_{\mathrm{ref}}(t) = h_{\mathrm{MF}}(-t)$.

This receiver correlates the received signal with the stored template. This is implemented as exemplified in Fig. 5.4. Firstly, the received signal is multiplied by the stored template

$$y_1(t) = u_{\mathrm{Rx}}(t) \cdot h_{\mathrm{ref}}(t) \tag{5.13}$$

and then the obtained signal $y_1(t)$ is integrated and sampled at $t = t_0$, namely

$$y(t_0) = \int_{t_0}^{t_0+T} y_1(t)dt = \int_{t_0}^{t_0+T} u_{\mathrm{Rx}}(t) \cdot h_{\mathrm{ref}}(t)dt \tag{5.14}$$

where the integral has been evaluated in the time interval $[t_0, t_0 + T]$, with T the duration of $u_{\mathrm{Rx}}(t)$.

---

[3]In the derivation of eq. (5.11), the following Fourier Transform properties are used

$$\begin{aligned} s(t - t_0) &= S(f)e^{-j2\pi f t_0} \\ s^*(-t) &= S^*(f) \end{aligned}$$

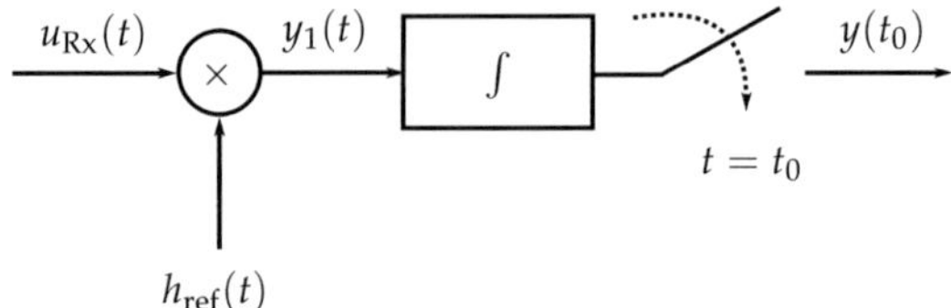

Figure 5.4: Block scheme of a correlation receiver.

It has to be pointed out that the previous equation can be rewritten as

$$\int_{t_0}^{t_0+T} u_{\mathrm{Rx}}(t) \cdot h_{\mathrm{ref}}(t)dt = \int_{t_0}^{t_0+T} u_{\mathrm{Rx}}(t) \cdot h_{\mathrm{ref}}(t+0)dt = R_{u_{\mathrm{Rx}}h_{\mathrm{ref}}}(0) \qquad (5.15)$$

i.e., the output of the correlation receiver represents the cross-correlation function $R_{u_{\mathrm{Rx}}h_{\mathrm{ref}}}(\tau)$ between the received signal $u_{\mathrm{Rx}}(t)$ and the template function $h_{\mathrm{ref}}(t)$ and this cross-correlation function is evaluated for the time difference $\tau = 0$.

In practical applications also multi-path propagation will occur, which will cause different pulse components arriving with different delays. However, if there is a line of sight propagation component, it can be assumed that the receiver will synchronize to the direct path and the delayed reflections will be outside the integration interval.

5.1.3.3 Evaluation of the SNR for the Matched Filter Receiver

Let suppose that the received waveform $x(t)$ consists of two terms (see Fig. 5.5): the signal $u_{\mathrm{Rx}}(t)$ received from a generic angular direction (θ, ψ) and an additive noise term $n_0(t)$. In the case of the matched filter, since the filtering operation performs a convolution, which is a linear operation, the obtained matched filter output $y(t)$ is given by

$$y(t) = (u_{\mathrm{Rx}}(t) + n_0(t)) * h_{\mathrm{MF}}(t) = y_r(t) + y_n(t) \qquad (5.16)$$

where $y_r(t) = u_{\mathrm{Rx}}(t) * h_{\mathrm{MF}}(t)$ is the signal component and $y_n(t) = n_0(t) * h_{\mathrm{MF}}(t)$ is the noise component after the matched filter. The SNR at the receiver output can be calculated as

$$SNR = \frac{Y_r}{Y_n} \qquad (5.17)$$

where Y_r is the signal power at the receiver output (i.e. of the signal component y_r) and Y_n is the noise power at the receiver output. Let N_0 be the noise power density before the receiver. It results[4]

$$Y_n = N_0 \int_{t_0}^{t_0+T} h_{\mathrm{MF}}^2(t)dt . \qquad (5.18)$$

---

[4]It has to be pointed out that since u_{Tx} has a finite duration T, from eq. (5.10) also h_{MF} has a finite duration T. Moreover, since all the considered time domain quantities are real, because baseband signals are regarded, for each considered signal s, it results that $|s| = s$.

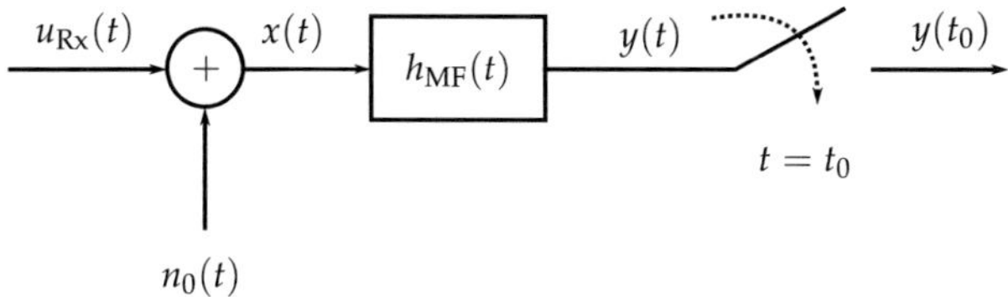

Figure 5.5: Block scheme of a matched filter receiver in presence of noise.

On the other hand, the signal power at the receiver output at the time $t = t_0$ can be calculated as

$$Y_r = y_r(t_0)^2 = \left[\int_{t_0}^{t_0+T} h_{\text{MF}}(t')u_{\text{Rx}}(t_0 - t')dt' \right]^2 . \tag{5.19}$$

Using eq. (5.18) and (5.19), from (5.17) the SNR can be written as (for $t = t_0$)

$$SNR = \frac{Y_r}{Y_n} = \frac{1}{N_0} \frac{\left[\int_{t_0}^{t_0+T} h_{\text{MF}}(t')u_{\text{Rx}}(t_0 - t')dt' \right]^2}{\int_{t_0}^{t_0+T} h_{\text{MF}}^2(t)dt} . \tag{5.20}$$

In order to simplify the previous equation and to transform it to an easier form, let recall that the received signal energy E is given by

$$E = \int_{t_0}^{t_0+T} u_{\text{Rx}}^2(t)dt = \int_{t_0}^{t_0+T} u_{\text{Rx}}^2(t_0 - t')dt' . \tag{5.21}$$

Hence, expanding eq. (5.20) by E, the SNR can be rewritten as

$$SNR = \frac{E}{N_0} \cdot \underbrace{\left(\frac{\left[\int_{t_0}^{t_0+T} h_{\text{MF}}(t')u_{\text{Rx}}(t_0 - t')dt' \right]^2}{\int_{t_0}^{t_0+T} h_{\text{MF}}^2(t)dt \cdot \int_{t_0}^{t_0+T} u_{\text{Rx}}^2(t_0 - t')dt'} \right)}_{\left(F_{u_{\text{Rx}}h_{\text{MF}}} \right)^2} = \frac{E}{N_0} \cdot \left(F_{u_{\text{Rx}}h_{\text{MF}}} \right)^2 . \tag{5.22}$$

The last term $F_{u_{\text{Rx}}h_{\text{MF}}}$ represents the normalized cross-correlation function between the received signal $u_{\text{Rx}}(t)$ and the matched filter impulse response $h_{\text{MF}}(t)$ for $t = t_0$. Hence, it can be concluded that the SNR is directly dependent on the cross-correlation function, for $t = t_0$, between the received signal and the matched filter impulse response.

Moreover, the maximum value of the SNR in the previous equation can be rewritten as

$$SNR_{\max} = \frac{E}{N_0} \max_{\tau} \frac{\left[\int_{t_0}^{t_0+T} h_{\text{MF}}(t')u_{\text{Rx}}(\tau - t')dt' \right]^2}{\int_{t_0}^{t_0+T} h_{\text{MF}}^2(t)dt \cdot \int_{t_0}^{t_0+T} u_{\text{Rx}}^2(\tau - t')dt'}$$

$$= \frac{E}{N_0} \cdot \max_{\tau} \left(F_{rh_{\text{MF}}}(\tau) \right)^2 \tag{5.23}$$

From the theory of the matched filter receiver, it can be concluded that the optimum value of τ that maximizes the SNR is exactly the time t_0 when the receiver template is matched to the received signal.

5.1.3.4 Evaluation of the *SNR* for the Correlation Receiver

Analogously to the previous case, let suppose that the received waveform is given by a signal component $u_{\text{Rx}}(t)$ and a noise component $n_0(t)$, as shown in Fig. 5.6. Also in this case, due to the linearity of the multiplication operation and of the integration with respect to the linear combination, the output of the receiver is given by

$$
y(t_0) = \int_{t_0}^{t_0+T} (u_{\text{Rx}}(t) + n_0(t)) \cdot h_{\text{ref}}(t)dt =
$$

$$
= \underbrace{\int_{t_0}^{t_0+T} u_{\text{Rx}}(t) \cdot h_{\text{ref}}(t)dt}_{y_r(t_0)} + \underbrace{\int_{t_0}^{t_0+T} n_0(t) \cdot h_{\text{ref}}(t)dt}_{y_n(t_0)} \tag{5.24}
$$

i.e. it is a summation of a signal term $y_r(t_0) = \int_{t_0}^{t_0+T} u_{\text{Rx}}(t) \cdot h_{\text{ref}}(t)dt$ and a term due to the noise $y_n(t_0) = \int_{t_0}^{t_0+T} n_0(t) \cdot h_{\text{ref}}(t)dt$. The power of the signal at the receiver output is

$$
Y_r = y_r(t_0)^2 = \left[\int_{t_0}^{t_0+T} u_{\text{Rx}}(t) \cdot h_{\text{ref}}(t)dt \right]^2 \tag{5.25}
$$

and the noise power at the receiver output can be calculated analogously to eq. (5.18), namely

$$
Y_n = N_0 \int_{t_0}^{t_0+T} h_{\text{ref}}^2(t)dt \ . \tag{5.26}
$$

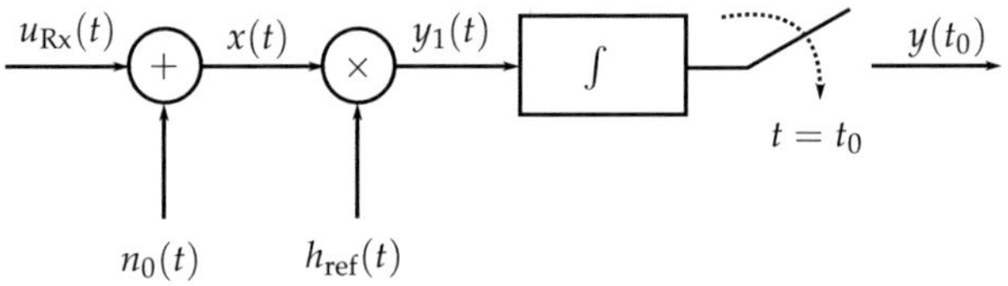

Figure 5.6: Block scheme of a correlation receiver in presence of noise.

Hence, applying eq. (5.17), the SNR is given by

$$
SNR = \frac{Y_r}{Y_n} = \frac{1}{N_0} \frac{\left[\int_{-\infty}^{\infty} u_{\text{Rx}}(t) h_{\text{ref}}(t)dt' \right]^2}{\int_{-\infty}^{\infty} h_{\text{ref}}^2(t)dt} \ . \tag{5.27}
$$

Also in this case, expanding the previous equation by the signal energy E derived in eq. (5.21) it gives

$$
SNR = \frac{E}{N_0} \frac{\left[\int_{-\infty}^{\infty} u_{\text{Rx}}(t) h_{\text{ref}}(t)dt \right]^2}{\int_{-\infty}^{\infty} h_{\text{ref}}^2(t)dt \cdot \int_{-\infty}^{\infty} u_{\text{Rx}}^2(t)dt} = \frac{E}{N_0} \cdot \left(F_{u_{\text{Rx}} h_{\text{ref}}} \right)^2 \ . \tag{5.28}
$$

By comparing eq. (5.22) and (5.28) it can be seen that in the case when the received signal and the template are matched, the implementation of the receiver through multiplication by a template is equivalent to the implementation by the matched filter. In that case, the SNR obtained through (5.28) is the maximum achievable and it coincides with (5.22). In the other cases, where the template is no more matched to the received signal, i.e. the received signal is no more impinging on the receive antenna from the main beam direction, the SNR is lower than the maximal achievable. In order to quantify its decrement, the term $F_{u_{\mathrm{Rx}}h_{\mathrm{ref}}}$ is now analyzed.

The signal term in eq. (5.24) can be rewritten as[5]

$$y_r(t) = \int_{-\infty}^{+\infty} u_{\mathrm{Rx}}(t) \cdot h_{\mathrm{ref}}(t)dt = u_{\mathrm{Rx}}(t) * h_{\mathrm{ref}}(-t) \tag{5.30}$$

i.e., the output value is given by the convolution between the received signal $u_{\mathrm{Rx}}(t)$ and a time-reversed version of the template function $h_{\mathrm{ref}}(-t)$. Using this observation, by considering the time domain link description from eq. (5.4), the term $F_{u_{\mathrm{Rx}}h_{\mathrm{ref}}}$ can be rewritten as (hereafter, only the numerator is considered, for simplicity, since the denominator is only a normalization and does not play any role in the developed analysis)

$$
\begin{aligned}
F_{u_{\mathrm{Rx}}h_{\mathrm{ref}}}(\theta_{\mathrm{Tx}}, \psi_{\mathrm{Tx}}, \theta_{\mathrm{Rx}}, \psi_{\mathrm{Rx}}) &= u_{\mathrm{Rx}}(t) * h_{\mathrm{ref}}(-t) \\
&= \left(\mathbf{h}_{\mathrm{Rx}}(t, \theta_{\mathrm{Rx}}, \psi_{\mathrm{Rx}}) * h_{\mathrm{Ch}}(t) * \mathbf{h}_{\mathrm{Tx}}(t, \theta_{\mathrm{Tx}}, \psi_{\mathrm{Tx}}) * \frac{\partial}{\partial t} u_{\mathrm{Tx}}(t) \right) * \\
&\quad \left(\mathbf{h}_{\mathrm{Rx}}^{\mathrm{M}}(-t, \theta_{\mathrm{Rx}}, \psi_{\mathrm{Rx}}) * h_{\mathrm{Ch}}^{\mathrm{id}}(-t) * \mathbf{h}_{\mathrm{Tx}}^{\mathrm{M}}(-t, \theta_{\mathrm{Tx}}, \psi_{\mathrm{Tx}}) * \frac{\partial}{\partial t} u_{\mathrm{Tx}}(-t) \right) \\
&= \underbrace{\mathbf{h}_{\mathrm{Rx}}(t, \theta_{\mathrm{Rx}}, \psi_{\mathrm{Rx}}) * \mathbf{h}_{\mathrm{Rx}}^{\mathrm{M}}(-t, \theta_{\mathrm{Rx}}, \psi_{\mathrm{Rx}})}_{R_{\mathbf{h}_{\mathrm{Rx}}\mathbf{h}_{\mathrm{Rx}}^{\mathrm{M}}}(0, \theta_{\mathrm{Rx}}, \psi_{\mathrm{Rx}})} * \underbrace{h_{\mathrm{Ch}}(t) * h_{\mathrm{Ch}}^{\mathrm{id}}(-t)}_{R_{h_{\mathrm{Ch}}h_{\mathrm{Ch}}^{\mathrm{id}}}(0)} * \\
&\quad \underbrace{\mathbf{h}_{\mathrm{Tx}}(t, \theta_{\mathrm{Tx}}, \psi_{\mathrm{Tx}}) * \mathbf{h}_{\mathrm{Tx}}^{\mathrm{M}}(-t, \theta_{\mathrm{Tx}}, \psi_{\mathrm{Tx}})}_{R_{\mathbf{h}_{\mathrm{Tx}}\mathbf{h}_{\mathrm{Tx}}^{\mathrm{M}}}(0, \theta_{\mathrm{Tx}}, \psi_{\mathrm{Tx}})} * \underbrace{\frac{\partial}{\partial t} u_{\mathrm{Tx}}(t) * \frac{\partial}{\partial t} u_{\mathrm{Tx}}(-t)}_{R_{\dot{u}_{\mathrm{Tx}}\dot{u}_{\mathrm{Tx}}}(0)}
\end{aligned}
\tag{5.31}
$$

where the commutativity property of the convolution has been used. The term

$$R_{\mathbf{h}_{\mathrm{Rx}}\mathbf{h}_{\mathrm{Rx}}^{\mathrm{M}}}(0, \theta_{\mathrm{Rx}}, \psi_{\mathrm{Rx}}) = \mathbf{h}_{\mathrm{Rx}}(t, \theta_{\mathrm{Rx}}, \psi_{\mathrm{Rx}}) * \mathbf{h}_{\mathrm{Rx}}^{\mathrm{M}}(-t, \theta_{\mathrm{M}}, \psi_{\mathrm{M}}) \tag{5.32}$$

represents the cross-correlation function between the impulse response of the receive antenna in a generic angular direction $\mathbf{h}_{\mathrm{Rx}}(t, \theta_{\mathrm{Rx}}, \psi_{\mathrm{Rx}})$ and in the main beam direction $\mathbf{h}_{\mathrm{Rx}}^{\mathrm{M}}(t, \theta_{\mathrm{M}}, \psi_{\mathrm{M}})$, and this cross-correlation function is evaluated at the time difference $\tau = 0$. Analogously,

$$R_{\mathbf{h}_{\mathrm{Tx}}\mathbf{h}_{\mathrm{Tx}}^{\mathrm{M}}}(0, \theta_{\mathrm{Tx}}, \psi_{\mathrm{Tx}}) = \mathbf{h}_{\mathrm{Tx}}(t, \theta_{\mathrm{Tx}}, \psi_{\mathrm{Tx}}) * \mathbf{h}_{\mathrm{Tx}}^{\mathrm{M}}(-t, \theta_{\mathrm{M}}, \psi_{\mathrm{M}}) \tag{5.33}$$

gives the cross-correlation function (evaluated for $\tau = 0$) between the impulse response of the transmit antenna in a generic angular direction $\mathbf{h}_{\mathrm{Tx}}(t, \theta_{\mathrm{Tx}}, \psi_{\mathrm{Tx}})$ and in the main

---

[5]In order to write (5.30), the following property of the cross-correlation between two signals $x(t)$ and $y(t)$ is used

$$R_{xy}(\tau) = \int_{-\infty}^{+\infty} x(t) \cdot y(t + \tau)dt = x(t) * y(-t) . \tag{5.29}$$

beam direction $\mathbf{h}_{\text{Tx}}^{\text{M}}(t, \theta_{\text{M}}, \psi_{\text{M}})$. Moreover, the term

$$R_{h_{\text{Ch}} h_{\text{Ch}}^{\text{id}}}(0) = h_{\text{Ch}}(t) * h_{\text{Ch}}^{\text{id}}(-t) \tag{5.34}$$

represents the cross-correlation function (evaluated for $\tau = 0$) between the channel impulse response and its idealized version included in the template. Finally, the term

$$R_{\dot{u}_{\text{Tx}} \dot{u}_{\text{Tx}}}(0) = \frac{\partial}{\partial t} u_{\text{Tx}}(t) * \frac{\partial}{\partial t} u_{\text{Tx}}(-t) = \dot{u}_{\text{Tx}}(t) * \dot{u}_{\text{Tx}}(-t) \tag{5.35}$$

gives the auto-correlation function (evaluated at $\tau = 0$) of the time derivative $\dot{u}_{\text{Tx}}(t)$ of the transmit signal $u_{\text{Tx}}(t)$.

It has to be pointed out that, since $\mathbf{h}_{\text{Tx}}(t, \theta_{\text{Tx}}, \psi_{\text{Tx}})$ and $\mathbf{h}_{\text{Rx}}(t, \theta_{\text{Rx}}, \psi_{\text{Rx}})$ are angular dependent, as seen in the previous chapter, the terms $R_{h_{\text{Rx}} h_{\text{Rx}}^{\text{M}}}(0, \theta_{\text{Rx}}, \psi_{\text{Rx}})$ and $R_{h_{\text{Rx}} h_{\text{Rx}}^{\text{M}}}(0, \theta_{\text{Tx}}, \psi_{\text{Tx}})$ are actually angular dependent.

Hence, from eq. (5.31), it can be concluded that in order to maximize the *SNR* in (5.28) in the case of a correlation receiver, the term $F_{u_{\text{Rx}} h_{\text{ref}}}(\theta_{\text{Tx}}, \psi_{\text{Tx}}, \theta_{\text{Rx}}, \psi_{\text{Rx}})$ has to be maximized. Since it is angular dependent, it has to be maximized depending on the angular directions $\theta_{\text{Tx}}, \psi_{\text{Tx}}, \theta_{\text{Rx}}, \psi_{\text{Rx}}$. From eq. (5.31), it can be recognized that, in order to maximize $F_{u_{\text{Rx}} h_{\text{ref}}}$, the single terms $R_{h_{\text{Rx}} h_{\text{Rx}}^{\text{M}}}(0, \theta_{\text{Rx}}, \psi_{\text{Rx}})$, $R_{h_{\text{Tx}} h_{\text{Tx}}^{\text{M}}}(0, \theta_{\text{Tx}}, \psi_{\text{Tx}})$, $R_{h_{\text{Ch}} h_{\text{Ch}}^{\text{id}}}(0)$ and $R_{\dot{u}_{\text{Tx}} \dot{u}_{\text{Tx}}}(0)$, from which it is composes, have to be singularly maximized. However, from eq. (5.34) and (5.35), the two terms $R_{h_{\text{Ch}} h_{\text{Ch}}^{\text{id}}}(0)$ and $R_{\dot{u}_{\text{Tx}} \dot{u}_{\text{Tx}}}(0)$ are fixed for a specific UWB radio link, i.e., they are not angular dependent and hence they cannot be maximized. On the other hand, from eq. (5.32) and (5.33) it can be evinced that the terms $R_{h_{\text{Rx}} h_{\text{Rx}}^{\text{M}}}(0, \theta_{\text{Rx}}, \psi_{\text{Rx}})$ and $R_{h_{\text{Tx}} h_{\text{Tx}}^{\text{M}}}(0, \theta_{\text{Tx}}, \psi_{\text{Tx}})$ are angular dependent and hence they have to be singularly maximized depending on the angular direction.

From the obtained results, it can be evinced that $F_{u_{\text{Rx}} h_{\text{ref}}}$ and hence the *SNR* is directly dependent on the cross-correlations functions, for $\tau = 0$, between the impulse response in the main radiation direction and the impulse response in the direction of the transmission link of the transmit antenna $R_{h_{\text{Tx}} h_{\text{Tx}}^{\text{M}}}(0, \theta_{\text{Tx}}, \psi_{\text{Tx}})$ and of the receive antenna $R_{h_{\text{Rx}} h_{\text{Rx}}^{\text{M}}}(0, \theta_{\text{Rx}}, \psi_{\text{Rx}})$. Since these cross-correlations are angular-dependent, also $F_{u_{\text{Rx}} h_{\text{ref}}}$ and hence the *SNR* are angular dependent. From these observations, it is important to have the possibility to quantify the angular variation of $R_{h_{\text{Tx}} h_{\text{Tx}}^{\text{M}}}(0, \theta_{\text{Tx}}, \psi_{\text{Tx}})$ and $R_{h_{\text{Rx}} h_{\text{Rx}}^{\text{M}}}(0, \theta_{\text{Rx}}, \psi_{\text{Rx}})$, which directly gives the angular variation of $F_{u_{\text{Rx}} h_{\text{ref}}}$ and hence the *SNR*.

In the following, the cross-correlation function of an antenna will be calculated (i.e. either the term in eq. (5.32) or (5.33)) and a criterion will be introduced, which permits to quantify the spatial regions where the radiated signal has optimum properties.

5.2 Quantification of the Antenna Angular Dependence

The quantification of the angular dependence of the signal radiated by the antenna is an important task for communication applications and Radar applications as well. In the communication link a variation of the signal in the different angular directions causes a decrement of the system performance, as previously derived (ref. to sec. 5.1.3.4). Hence, having the possibility of quantifying the signal variation in the different angular directions, it is also possible to estimate the variation of the SNR for impulse radio transmission due to this non-ideal antenna behavior.

On the other hand, in Radar applications, the variation of the received signal with respect to the transmitted signal contains the target signature. Hence, an angular signal variation due to the non-ideal antenna behavior can introduce erroneous target information or affect the actually expected one. Consequently, it is important to have the possibility of knowing a priori the variation which the transmit signal undergoes, due to only the antenna non-ideal behavior. From that knowledge it is possible to discern the signal distortion due to the actual target information from the signal distortion due to the antenna angular dependent behavior and to limit the operation of the Radar to an angular region with negligible variation of the impulse response. The analysis in the Radar case is one topic of chapter 7.

In literature [67] an analysis of the distortion of the radiated pulse due to a variation of the antenna gain or group delay is reported. In other references [9], [10], the cross-correlation is used to investigate the behavior with respect to different radiated pulses. In [11] measurements of the cross-correlation between transmit and receive antenna arrays are presented. However, these measurements are only shown depending on the distance between the transmit and the receive arrays and the number of elements of the arrays themselves for a particular scenario, without taking into consideration the three-dimensional case and the different angular directions. Hence, from that analysis it is not possible to assess the effect of the distortion introduced by the antenna in the different angular directions. In [68] the correlation between a transmit signal and a template is defined and is used in an optimization problem. However, also in this case, the correlation is presented as a function of the transmit signal and not as a measure of the intrinsic properties of the antennas.

All these investigations do not allow to classify the spatial variation of the antenna characteristics. A suitable approach for that purpose will be derived in the following.

5.2.1 Fidelity Analysis

The aim is to introduce criteria for the single antenna (either Tx or Rx) in order to judge its angular dependent behavior. As it as been seen in section 5.1, the cross-correlation function, between the impulse response in the main radiation direction and a generic angular direction, is an important criterion.

However, for a complete assessment, also the gain of the antenna must be considered. In the following, both aspects will be regarded for the field radiated from a single antenna.

The variation, which the transmitted signal undergoes in a generic angular direction with respect to the signal transmitted in the antenna's main beam direction, due to the angular dependence of the antenna properties, can be quantified through the spatial analysis of the cross-correlation function between the radiated signal in a generic angular direction (θ, ψ), and the signal transmitted in the main radiation direction (θ_M, ψ_M). Consequently, if a spatial correlation pattern is constructed, from its analysis it is possible to determine the size of the angular regions where the signal is preserved, i.e. where the distortion effects of the antenna are low with respect to the signal in the main beam direction.

According to the UWB radio link equation previously derived (see sec. 5.1.1), the radiated field in the main beam direction (θ_M, ψ_M) at a distance r from the transmit antenna itself is

$$\mathbf{e}_{Tx}^{M}(t, \theta_M, \psi_M, r) = \frac{K}{r}\delta\left(t - \frac{r}{c_0}\right) * h_{Tx}^{M}(t, \theta_M, \psi_M) * \frac{\partial}{\partial t}u_{Tx}(t) \qquad (5.36)$$

where the factor $K = \frac{\sqrt{Z_0}}{\sqrt{Z_{Tx}}} \cdot \frac{1}{2\pi c_0}$ and the antenna's angular dependence has been explicitly written.

Hence, as introduced in chapter 2, calculating the cross-correlation function between the transmitted signal $\mathbf{e}_{Tx}^{M}$ in the main direction (θ_M, ψ_M) and the transmitted signal $\mathbf{e}_{Tx}$ in the generic angular direction (θ, ψ) (see eq. (5.1)), gives a quantification of the distortion introduced by the single antenna in the different angular directions. According to the fidelity analysis introduced in chapter 2, in the analyzed case the fidelity term results

$$F(\theta, \psi, \tau) = \frac{R_{\mathbf{e}_{Tx}\mathbf{e}_{Tx}^{M}}(\theta, \psi, \tau)}{\sqrt{\int_{t_0}^{t_0+T}||\mathbf{e}_{Tx}||_2^2}\sqrt{\int_{t_0}^{t_0+T}||\mathbf{e}_{Tx}^{M}||_2^2}} = \frac{\mathbf{e}_{Tx}(\theta, \psi, t) * \mathbf{e}_{Tx}^{M}(\theta_M, \psi_M, \tau - t)}{\sqrt{\int_{t_0}^{t_0+T}||\mathbf{e}_{Tx}||_2^2}\sqrt{\int_{t_0}^{t_0+T}||\mathbf{e}_{Tx}^{M}||_2^2}}. \qquad (5.37)$$

Expanding the nominator in the previous equation, it gives

$$R_{\mathbf{e}_{Tx}\mathbf{e}_{Tx}^{M}}(\theta, \psi, \tau) = \left(h_{Ch}(t) * \mathbf{h}_{Tx}(t, \theta, \psi) * \dot{u}_{Tx}(t)\right) *$$

$$\left(h_{Ch}^{id}(\tau - t) * \mathbf{h}_{Tx}^{M}(\tau - t, \theta_M, \psi_M) * \dot{u}_{Tx}(\tau - t)\right)$$

$$= \underbrace{h_{Ch}(t) * h_{Ch}^{id}(\tau - t)}_{R_{h_{Ch}h_{Ch}}(\tau)} * \underbrace{\mathbf{h}_{Tx}(t, \theta, \psi) * \mathbf{h}_{Tx}^{M}(\tau - t, \theta_M, \psi_M)}_{R_{\mathbf{h}_{Tx}\mathbf{h}_{Tx}^{M}}(\tau, \theta, \psi)}$$

$$* \underbrace{\dot{u}_{Tx}(t) * \dot{u}_{Tx}(\tau - t)}_{R_{\dot{u}_{Tx}\dot{u}_{Tx}(t)}(\tau)} \qquad (5.38)$$

Comparing the previous equation with eq. (5.31), it can be recognized that (5.38) is composed by three terms: the first one is the auto-correlation of the channel $R_{h_{Ch}h_{Ch}}(\tau)$, evaluated at the time difference τ (ref. to eq. (5.34)). The second term is the cross-correlation function $R_{\mathbf{h}_{Tx}\mathbf{h}_{Tx}^{M}}(\tau, \theta, \psi)$ between the impulse response of the transmit antenna in the main beam direction and in a generic direction, evaluated at the time difference τ (ref. eq. (5.33)). The last term is the auto-correlation function of the time derivative of the transmit signal $\dot{u}_{Tx}$, evaluated at the time difference τ (ref. to (5.35)). Since it has been assumed that the channel is a delay channel, according to eq. (5.2), this term performs only a shift of the pulses and an attenuation of a factor $1/r$ depending on the distance, but these operations do not influence the cross-correlation calculation[6]. Hence, also in this case it can be evinced that the knowledge of the cross-correlation properties of the antennas has an important role, since it permits to quantify the distortion, which the signal undergoes. The transmitted signal is minimally distorted with respect to the signal transmitted in the main beam direction when the fidelity term (5.37) is maximal (ref. to chapter 2), i.e. for

$$F_{\max}(\theta, \psi) = \max_{\tau} F(\theta, \psi, \tau) \ . \tag{5.39}$$

From these considerations it can be concluded that the fidelity factor has an important meaning in the analysis of UWB antennas. From its knowledge the worsening of the SNR can directly be evaluated, according to eq. (5.28). Moreover, it also quantifies the angular dependent distortion operated by the antenna on the transmitted signal.

5.2.2 Establishment of a joint Criterion

As already pointed out in chapter 2, the fidelity factor does not characterize the antenna's gain. Hence, also the peak of the impulse response $P(\theta, \psi)$ in the different angular directions has to be considered. A good antenna performance requires both, a high fidelity/cross-correlation and high peak of the impulse response simultaneously. Consequently, a criterion which permits to identify the area where the radiated signal is well correlated with the signal in the main direction and preserves a high peak is needed.

From the analysis of the peak in the different angular directions it is possible to determine the areas where the antenna radiates high peak power. A good time domain behavior of an antenna is given in areas where the radiated signal is well correlated with the signal in the main beam direction (low distortion) and also the radiated peak power is high (high peak of the transient response). Hence, in order to determine such areas, a joint criterion has to be established.

For that reason, the product of the absolute value of the fidelity $|F|$ and the peak P for each angular direction (θ_i, ψ_i) is regarded, normalized to its maximum, namely

$$PF(\theta, \psi) = \frac{P(\theta, \psi) \cdot |F(\theta, \psi)|}{\max_{\theta, \psi} \{P(\theta, \psi) \cdot |F(\theta, \psi)|\}} \ . \tag{5.40}$$

By this procedure in each direction the peak value P is weighted with the fidelity $|F|$. Hence, the obtained area with high values is the area where the radiated signal is well correlated with the signal in the main beam direction and also has a high peak power. Consequently, in that area the antenna has a high performance for impulse radio operations.

---

[6]In fact, the cross-correlation function evaluates the shape of the signals and in this case both the signals in the main direction and in the generic direction are evaluated at a distance r and consequently equally weighted by the channel influence.

5.2.3 Evaluation of the Fidelity from Measurement Results

In order to calculate the fidelity term in eq. (5.37) and its maximal value (see eq. (5.39)), the knowledge of the antenna impulse response is necessary. As seen in the previous chapter (ref. to eq. (4.2) and sections 4.1.1.1 and 4.1.2), the impulse response of an antenna under test is calculated for a particular polarization of the reference receive antenna. Hereafter the horizontal $^\mathrm{h}$ and vertical $^\mathrm{v}$ polarizations of the receive antenna will be considered. From measurements performed according to sections 4.1.1.1 and 4.1.2, the impulse response $h^\mathrm{h}(t,\theta,\psi)$ of the antenna under test for the horizontal polarization of the receiver and $h^\mathrm{v}(t,\theta,\psi)$ of the antenna under test for the vertical polarization of the receiver are separately recovered. For each polarization, the maximum of the fidelity term is than calculated, according to eq. (5.39), obtaining $F_{\max}^\mathrm{h}(\theta,\psi)$ for the horizontal polarization of the receive antenna and $F_{\max}^\mathrm{v}(\theta,\psi)$ for the vertical polarization of the receive antenna.

It has to be observed that, for practical performance evaluations both polarization components must be jointly regarded.

This is done in the following regarding, for each particular angular direction (θ,ψ), the maximum value between the absolute value of the fidelity of the antenna impulse response for the vertical receiver polarization $|F_{\max}^\mathrm{v}(\theta,\psi)|$ and the absolute value of the fidelity of the antenna impulse response for the horizontal receiver polarization $|F_{\max}^\mathrm{h}(\theta,\psi)|$ in that angular direction, namely

$$F(\theta,\psi) = \max_{\theta,\psi}\left[|F_{\max}^\mathrm{v}(\theta,\psi)|, |F_{\max}^\mathrm{h}(\theta,\psi)|\right] . \tag{5.41}$$

In order to evaluate the overall antenna performance, also the peak P of the envelope of the impulse response of the antenna under test is investigated. In order to consider both polarizations, the antenna impulse response $h^{\mathrm{h+v}}$, given by the vector sum of the impulse responses of the vertical and horizontal receiver polarizations, namely

$$h^{\mathrm{h+v}}(t,\theta,\psi) = \sqrt{\left(h^\mathrm{h}(t,\theta,\psi)\right)^2 + \left(h^\mathrm{v}(t,\theta,\psi)\right)^2} \tag{5.42}$$

is considered. The angular, (θ,ψ)-dependent peak P is then calculated using eq. (2.7).

In order to evaluate the criterion introduced in eq. (5.40), the fidelity term obtained from eq. (5.41) is multiplied by the peak P calculated from the impulse response of eq. (5.42) and then normalized to its maximum. This is done for each angular direction (θ,ψ).

5.3 Analysis of UWB Antennas

In this section, the theoretical analysis and the criteria previously introduced are applied to particular antennas and supported by measurements. In the following the investigated antennas are presented together with the measurement setup.

5.3.1 Investigated Antennas

The analysis will be focused on two commonly used UWB antennas with different radiation principles, a traveling wave antenna (Vivaldi antenna) and a small antenna (Bow-tie antenna), in order to investigate and to compare their behaviors regarding their applicability for different applications.

5.3.1.1 Vivaldi Antenna

The Vivaldi antenna [6] is fed by aperture coupling and optimized for the frequency range from 3.1 GHz to 10.6 GHz. Its dimensions are 78 mm $\times$ 75 mm and it is shown in Fig. 5.7. The absolute value of the measured S_{11} parameter of the investigated Vivaldi antenna is plotted in Fig. 5.8. The antenna shows good input impedance matching from 2 to 12 GHz.

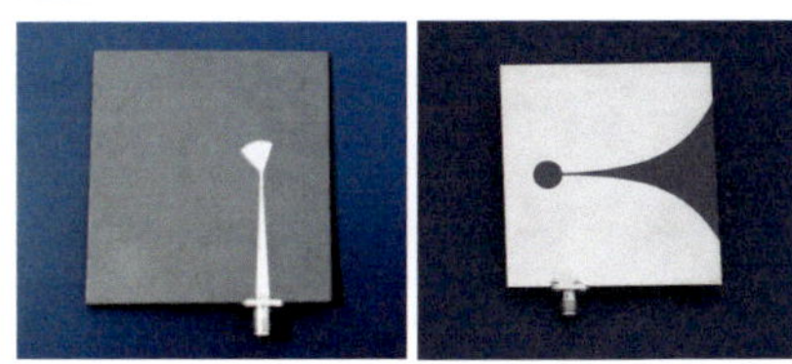

Figure 5.7: The investigated Vivaldi antenna: rear side (left) and front side (right).

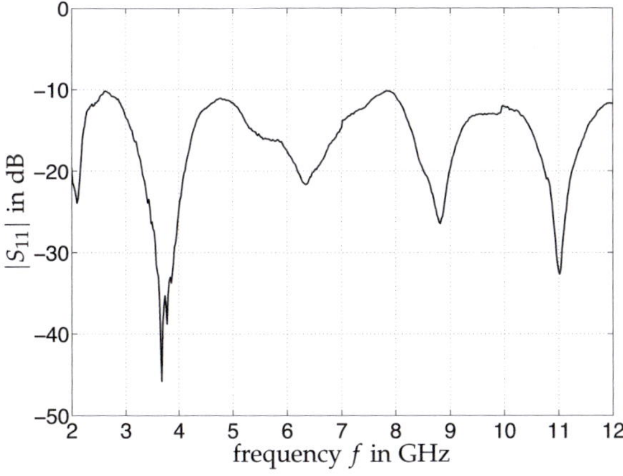

Figure 5.8: Measured $|S_{11}|$ parameter of the investigated Vivaldi antenna.

5.3.1.2 Bow-tie Antenna

The Bow-tie antenna [6] is fed by aperture coupling and optimized for the frequency range from 3.1 GHz to 10.6 GHz. Its dimensions are 36 mm $\times$ 31 mm. The antenna is shown in Fig. 5.9. In Fig. 5.10 the absolute value of the measured S_{11} parameter is plotted. In the frequency range from 3 to 10 GHz a good impedance matching can be observed.

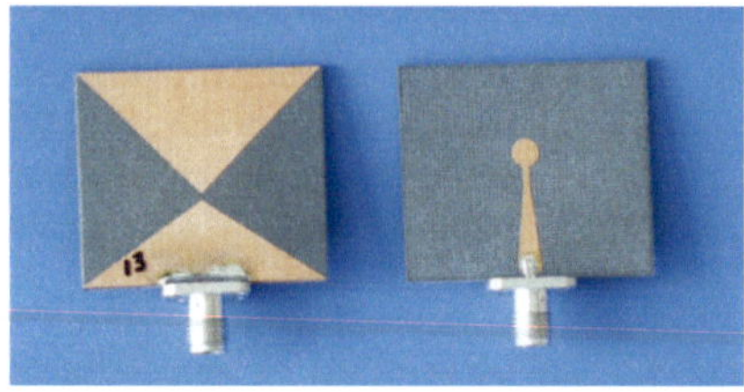

Figure 5.9: The investigated Bow-tie antenna: front side (left) and rear side (right).

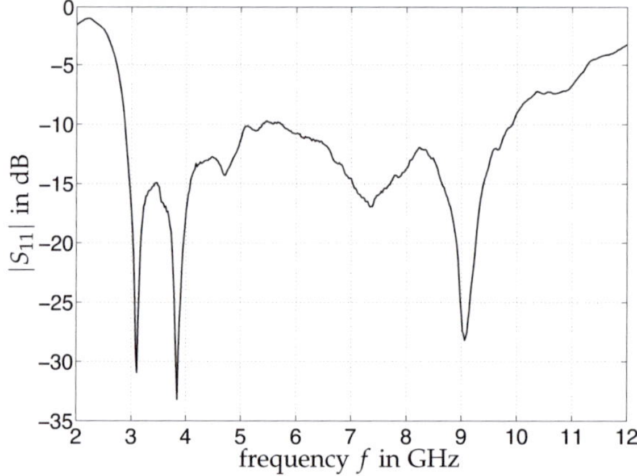

Figure 5.10: Measured $|S_{11}|$ parameter of the investigated Bow-tie antenna.

5.3.2 Measurement Setup

The AUT was placed in the center of a spherical anechoic chamber (JRC Ispra, diameter 20 meters) on a rotating tower. The azimuth angle was controlled by rotating the tower. The receiving antenna is a dual polarized horn antenna mounted on a carriage that moves along a rail to any desired elevation angle. The entire upper hemisphere was measured with an angular resolution of 5 degrees in both azimuth and elevation for both, the vertical and the horizontal polarization. For each position with a vector network analyzer the transfer coefficients were measured from 2 GHz to 18 GHz at 801 equidistantly spaced frequencies for both horizontal and vertical polarization simultaneously. For the calibration of the setup a direct through-connection measurement of the cables at the antenna feed-points and a transmission measurement between two identical horn antennas were taken [69]. Fig. 5.11 shows the coordinate system.

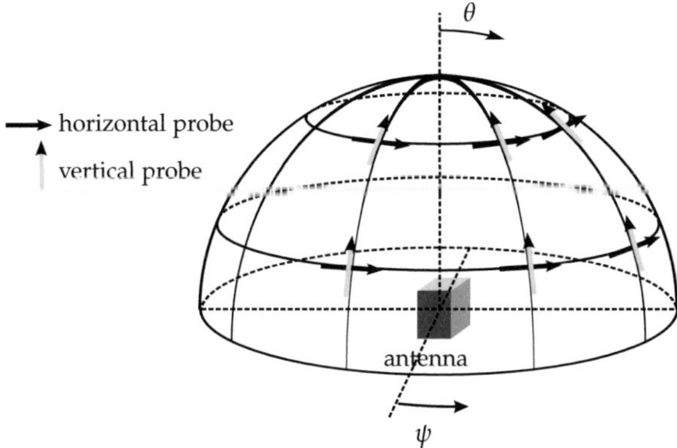

Figure 5.11: Coordinates for the measurement; gray: vertical polarization; black: horizontal polarization.

The resulting measured transfer function is obtained with a resolution of $\Delta f = 19.97$ MHz. Then, applying the procedure explained in section 2.2, the antenna transient response has been calculated through a discrete Fourier Transform starting from the analytical transfer function H^+_{AUT} (ref. to Appendix A.1), namely

$$h_{\mathrm{AUT}}(k\Delta t) = \Re\left[\frac{1}{N\Delta t}\sum_{n=0}^{N-1} H^+_{\mathrm{AUT}}(n\Delta f)\cdot\exp\left[j\frac{2\pi}{N}\right]\right]. \tag{5.43}$$

In the previous equation, the frequency range from 0 to 2 GHz has been filled with zeros. In order to improve the time domain accuracy, a zero padding has been applied at the end of length ten times the length of the measured data. The interpolated signal has an accuracy of $\Delta t = 5.5$ ps.

5.3.3 Obtained Results

In the following, the results for the correlation of the impulse response shapes and for the joint criterion regarding also the pulse peak of the two investigated antennas are presented.

5.3.3.1 Vivaldi Antenna

The Vivaldi antenna was placed in the anechoic chamber with respect to the coordinate system as illustrated in Fig. 5.12. The antenna impulse response has been plotted for different angular directions in Fig. 5.13 for horizontal polarization of the receiver (where the angular direction $(\theta, \psi) = (0°, 90°)$ corresponds to co-polarization), and in Fig. 5.14 for vertical polarization of the receiver (where the direction $(\theta, \psi) = (0°, 0°)$ corresponds to co-polarization), showing its angular dependence.

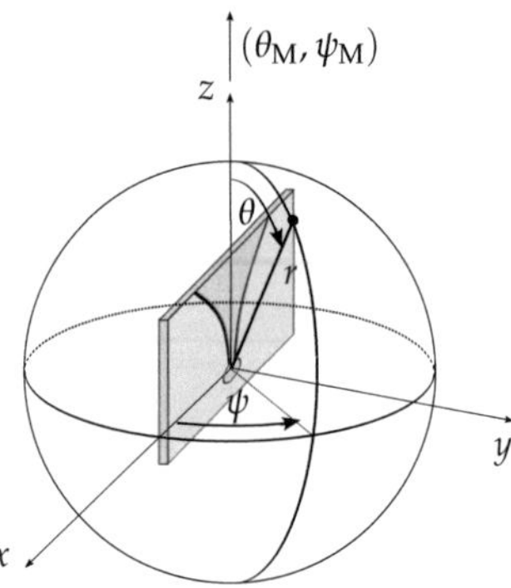

Figure 5.12: Position of the Vivaldi antenna with respect to the coordinate system and highlight of the main beam direction (θ_M, ψ_M).

In the following, the results for the fidelity for the Vivaldi antennas are presented. At the receiver side, both horizontal and vertical polarization components have been recovered separately, as previously discussed in section 5.2.3. When both polarization components are considered, the obtained results are presented in the following. Fig. 5.15 shows for each angular direction (θ, ψ), the maximum value between the absolute value of the fidelity of the antenna transient response for the vertical receiver polarization $|F^v_{max}(\theta, \psi)|$ and the absolute value of the fidelity of the antenna transient response for the horizontal receiver polarization $|F^h_{max}(\theta, \psi)|$ in that angular direction. It has been calculated according to eq. (5.41). The obtained results are plotted in a 2D projection of the upper hemisphere. The orientation of the polarization components is shown in Fig. 5.11. Each point corresponds to a particular (θ, ψ) direction. The ψ angles increase anticlockwise, while the θ angles increase from $0°$ at the center of the plot to $90°$ at the outer border of the picture. In order to better understand the antenna behavior, also a schematic representation of the antenna itself is plotted, showing its position and orientation in the anechoic chamber (gray lines represent the metallization surfaces of the antenna).

The fidelity has an high absolute value ($|F(\theta, \psi)| > 0.8$) in an elliptic angular area, for all ψ directions, comprised between $0° < \theta < 60°$.

In order to evaluate the overall antenna performance, also the peak P of the envelope of the Vivaldi antenna impulse response is investigated. In order to consider both polarizations, the antenna impulse response $h^{h+v}(t, \theta, \psi)$, given by eq. (5.42), is regarded.

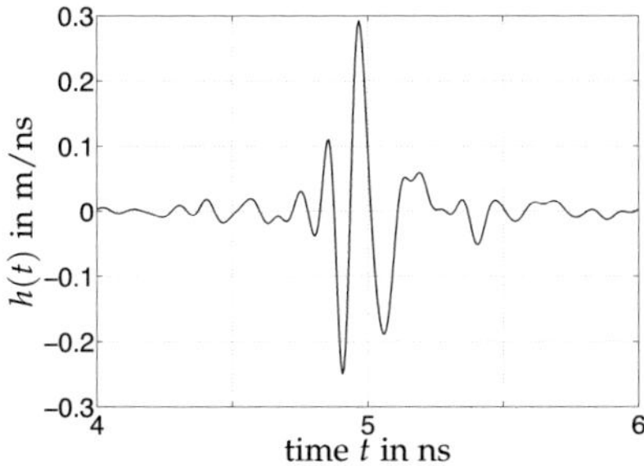

(a) horizontal polarization of the receive antenna:
$\theta = 0°$, $\psi = 90°$

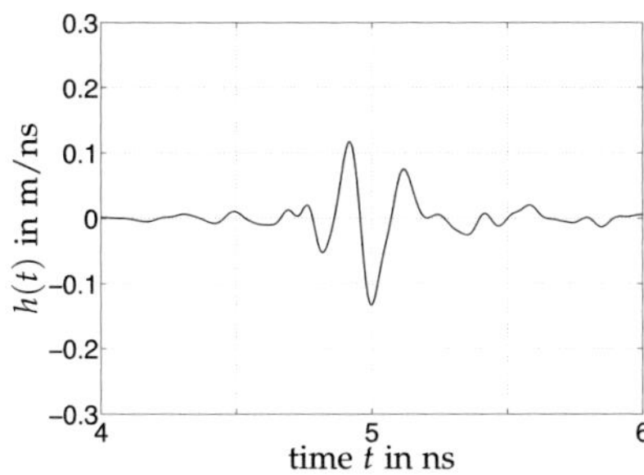

(b) horizontal polarization of the receive antenna:
$\theta = 45°$, $\psi = 90°$

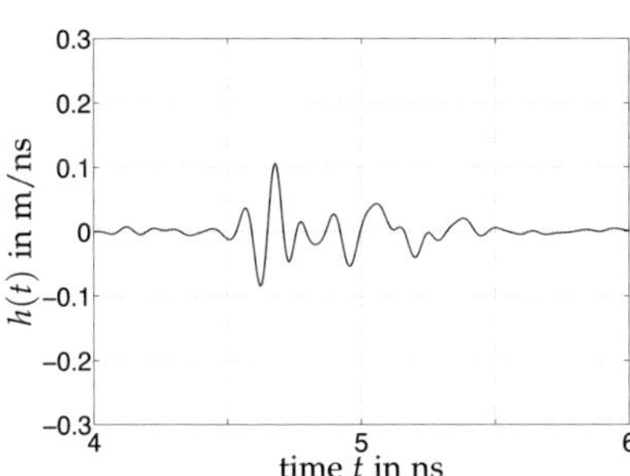

(c) horizontal polarization of the receive antenna:
$\theta = 90°$, $\psi = 90°$

Figure 5.13: The Vivaldi antenna impulse response in various angular directions for horizontal polarization of the receiver antenna (the direction $(\theta, \psi) = (0°, 90°)$ corresponds to co-polarization).

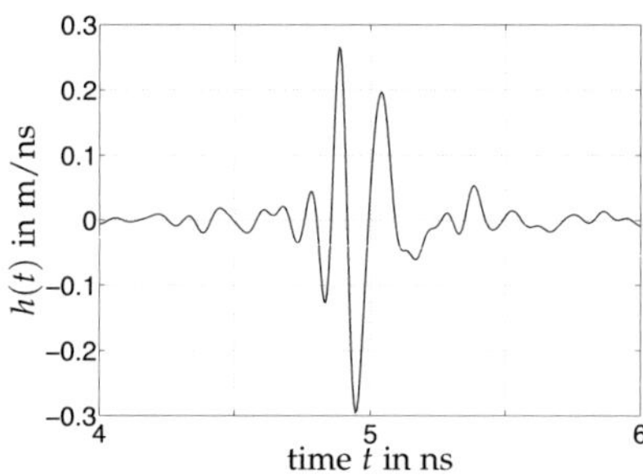

(a) vertical polarization of the receive antenna:
$\theta = 0°$, $\psi = 0°$

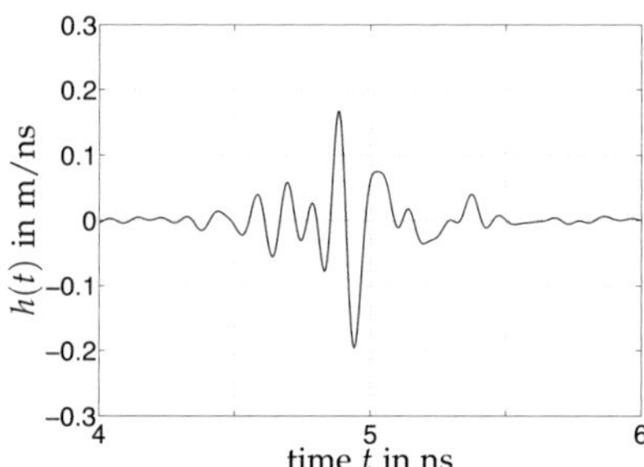

(b) vertical polarization of the receive antenna:
$\theta = 45°$, $\psi = 0°$

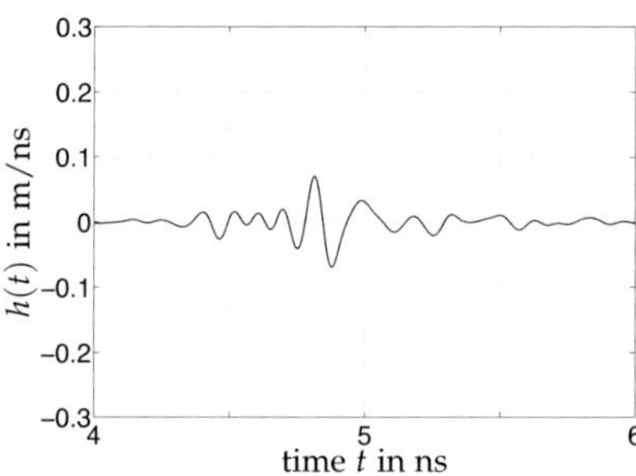

(c) vertical polarization of the receive antenna:
$\theta = 90°$, $\psi = 0°$

Figure 5.14: The Vivaldi antenna impulse response in various angular directions for vertical polarization of the receiver antenna (the direction $(\theta, \psi) = (0°, 0°)$ corresponds to co-polarization).

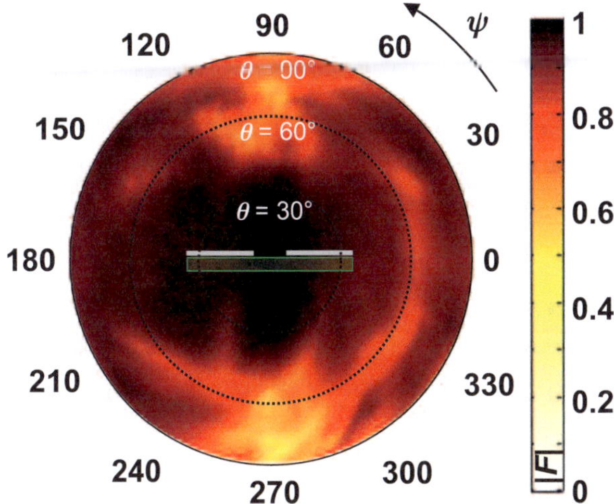

Figure 5.15: Spherical fidelity pattern $|F(\theta, \psi)|$ with respect to the main beam direction of the Vivaldi antenna for both polarizations of the receiver.

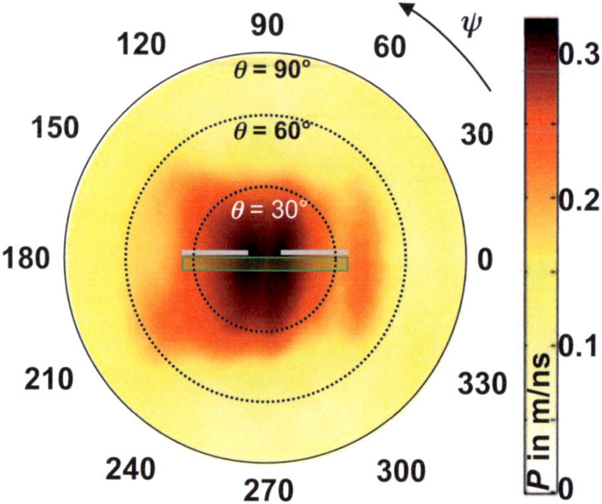

Figure 5.16: Peak $P(\theta, \psi)$ of the Vivaldi antenna transient response for both polarizations of the receiver.

The angular (θ, ψ) dependent peak P is then calculated using eq. (2.7). The obtained results are plotted in Fig. 5.16. The peak is high ($P > 0.25$ m/ns) in an elliptic angular area for $0° < \theta < 30°$ and every direction ψ. The transmitted signal presents its highest absolute values in the angular regions where it is also well correlated with the signal in the main radiation direction. This means that in the region where the signal has high gain the fidelity is also high.

In order to find the area where the antenna has a good behavior, i.e. where the radiated signal is well correlated with the signal in the main beam direction (low distortion) and also the radiated peak power is high, the criterion introduced in section 5.2.2 is now applied. The obtained result (normalized to its maximum) is plotted in Fig. 5.17. The area where the signal is weakly distorted and the radiated energy is high is in an elliptic area comprised between $0° < \theta < 30°$ and every direction ψ.

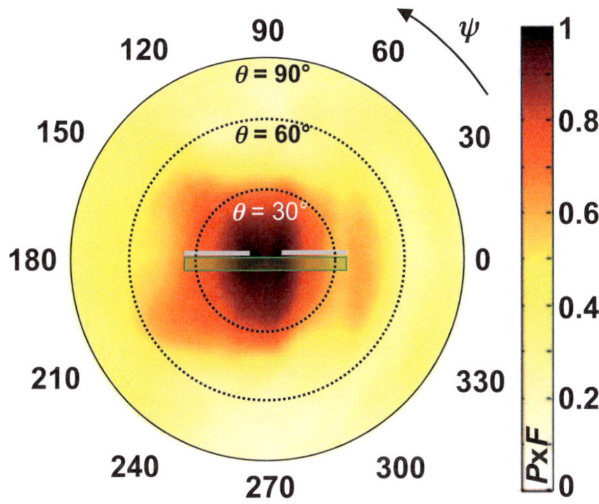

Figure 5.17: Normalized result for the joint performance criterion for the Vivaldi antenna.

5.3.3.2 Bow-tie Antenna

In Fig. 5.18 the position in the anechoic chamber of the Bow-tie antenna with respect to the coordinate system is shown. In Fig. 5.19 the measured impulse responses for the horizontal polarization of the receiver for different angular directions are plotted (the direction $(\theta, \psi) = (90°, 270°)$ corresponds to co-polarization). In Fig. 5.20 the measured impulse responses for the vertical polarization of the receiver for different angular directions are plotted (the direction $(\theta, \psi) = (90°, 270°)$ corresponds to co-polarization). It can be seen that in the case of the vertical polarization the antenna transient response does not change significantly for the directions $(\theta, \psi) = (90°, 270°)$, $(90°, 225°)$, $(90°, 180°)$, showing the almost omni-directionality of the Bow-tie antenna in the azimuth plane, as it discussed in the following. On the other hand, the horizontal polarization component changes highly with the angular directions (Fig. 5.19).

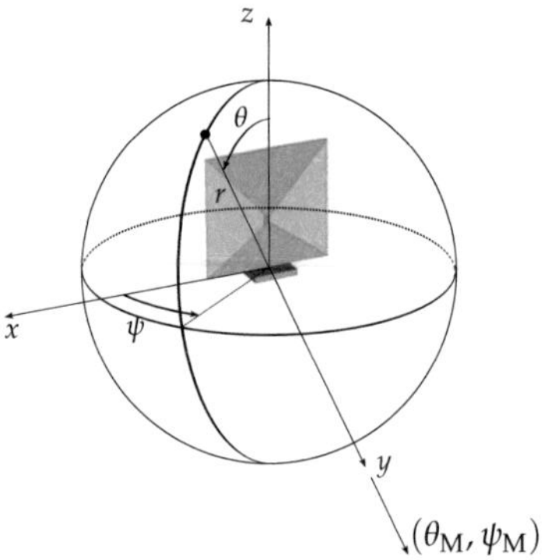

Figure 5.18: Position of the Bow-tie with respect to the coordinate system and highlight main beam direction (θ_M, ψ_M).

Also for the Bow-tie antenna, as for the Vivaldi antenna, from the measured data, the antenna transient responses for the vertical and the horizontal receiver polarizations are recovered separately, as discussed in section 5.2.3. Then, the fidelity patterns for the horizontal polarization of the receiver, $F^h_{max}(\theta, \psi)$, and for the vertical polarization of the receiver, $F^h_{max}(\theta, \psi)$, with respect to the main beam direction (shown in Fig. 5.18), are calculated using eq. (5.39). Once these fidelity terms are computed, through eq. (5.41) both polarizations are considered together. The obtained result for the fidelity is reported in Fig. 5.21. Here, also a schematic representation of the antenna itself is plotted, showing its position and orientation in the anechoic chamber (dark lines represent the metallization surfaces of the antenna).

The antenna presents high correlation $(|F(\theta, \psi)| > 0.8)$ in an angular area between $60° < \theta < 90°$ and all directions ψ. In this picture it is possible to recognize the omni-directionality of the Bow-tie antenna in the equatorial plane. Together with the fidelity F, also the peak P is investigated. It is calculated firstly constructing the antenna transient response $h^{h+v}(t, \theta, \psi)$, according to eq. (5.42), and than applying to it eq. (2.7) for calculating the peak P. The obtained results are plotted in Fig. 5.21.

It is possible to recognize the almost omni-directionality of the Bow-tie antenna in the equatorial plane since it presents a high peak for all ψ and for $60° < \theta < 90°$.

Finally, in order to identify the area where the Bow-tie antenna has good behavior (low distortion and high radiated peak power) the introduced joint criterion has been applied. The results are shown in Fig. 5.23, normalized to the maximum. The antenna

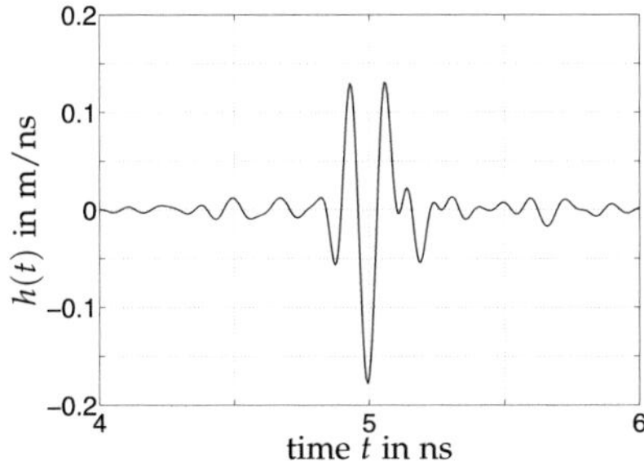

(a) horizontal polarization of the receive antenna:
$\theta = 90°$, $\psi = 270°$

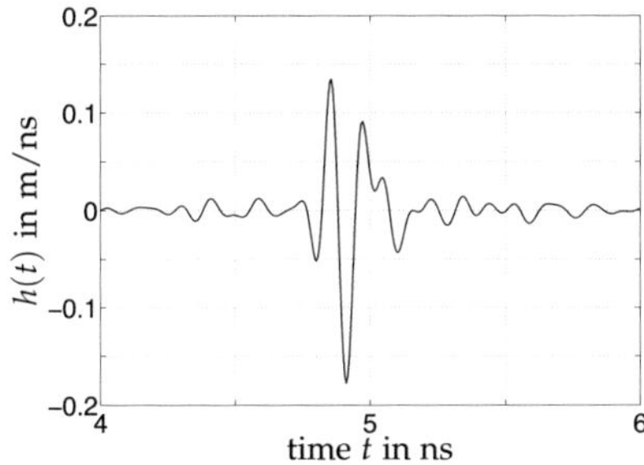

(b) horizontal polarization of the receive antenna:
$\theta = 45°$, $\psi = 270°$

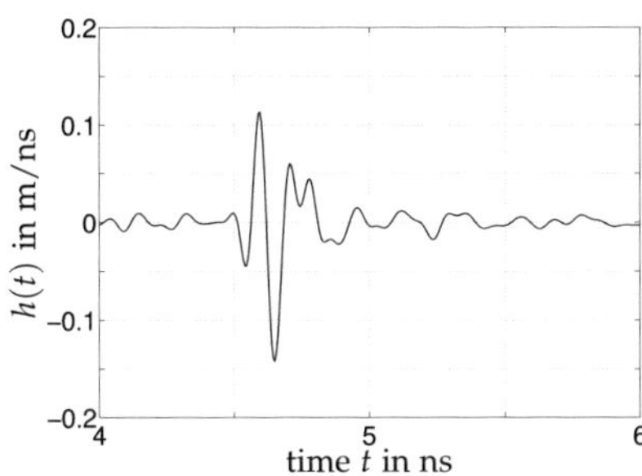

(c) horizontal polarization of the receive antenna:
$\theta = 0°$, $\psi = 270°$

Figure 5.19: The Bow-tie antenna impulse response in various angular directions for horizontal polarization of the receiver antenna (the direction $(\theta, \psi) = (90°, 270°)$ corresponds to co-polarization).

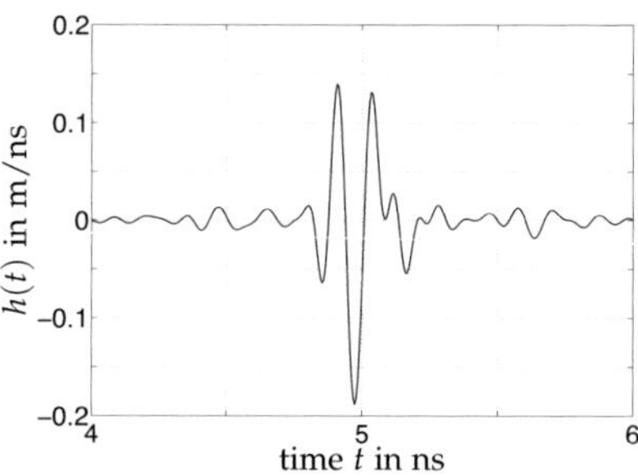

(a) vertical polarization of the receive antenna: $\theta = 90°$, $\psi = 270°$

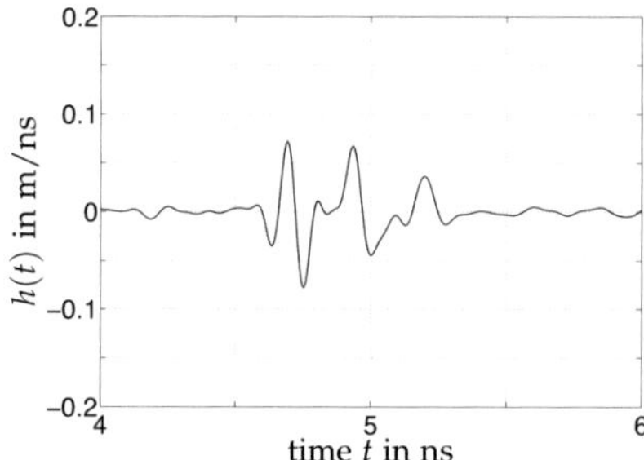

(b) vertical polarization of the receive antenna: $\theta = 90°$, $\psi = 225°$

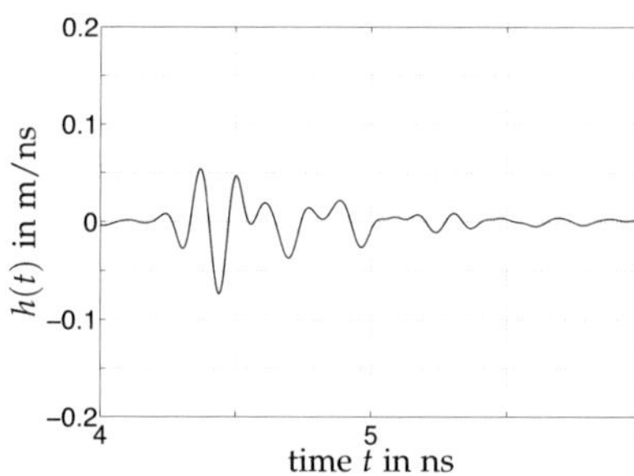

(c) vertical polarization of the receive antenna: $\theta = 90°$, $\psi = 180°$

Figure 5.20: The Bow-tie antenna impulse response in various angular directions for vertical polarization of the receiver antenna (the direction $(\theta, \psi) = (90°, 270°)$ corresponds to co-polarization).

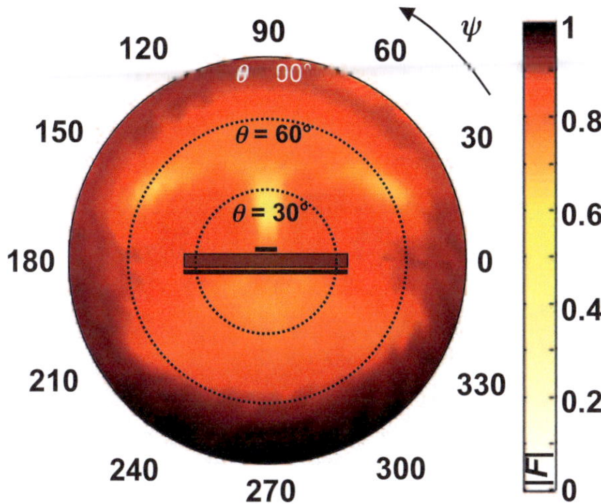

Figure 5.21: Spherical fidelity pattern $|F(\theta, \psi)|$ with respect to the main beam direction of the Bow-tie antenna for both polarizations of the receiver.

has good time domain behavior in all directions ψ for $60° < \theta < 90°$. Moreover, the two areas in proximity of $\psi = 0°$ and $\psi = 180°$ do not present high values. Even if these areas present average level of fidelity ($|F| < 0.7$) the antenna does not have optimal behavior in these areas since the peak is not very high ($P < 0.13$ m/ns). Hence, the angular region where the Bow-tie antenna preserves the transmitted pulse and radiates a pulse which has a high peak power is the equatorial region for $60° < \theta < 90°$ for all angular directions ψ.

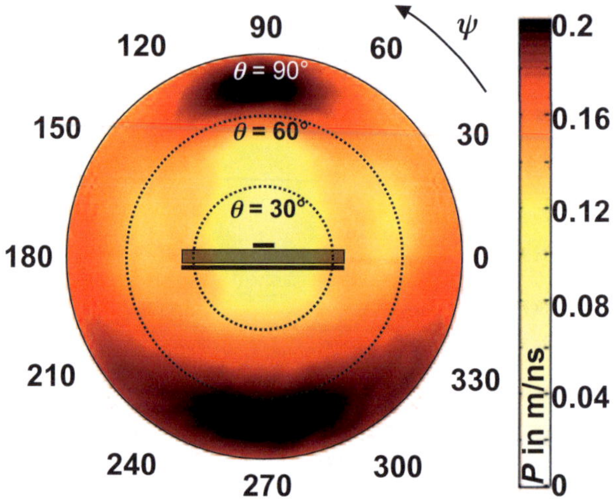

Figure 5.22: Peak $P(\theta, \psi)$ of the Bow-tie antenna transient response for both polarizations of the receiver.

5.4 Conclusion

In this chapter an analysis of the UWB radio link has been presented. In this analysis, firstly the mathematical description of the UWB radio link in the time domain and in the frequency domain has been derived. Secondly, the impact of the non-ideal system behavior (in particular the non-idealities due to the antennas and their angular dependence) on the system performance has been quantified. This has been done by evaluating the SNR at the receiver side in the case of a correlation receiver. It has been found out that the SNR is directly influenced by the angular-dependent behavior of the antennas. Hence, it has been seen that in order to quantify the variation of the SNR

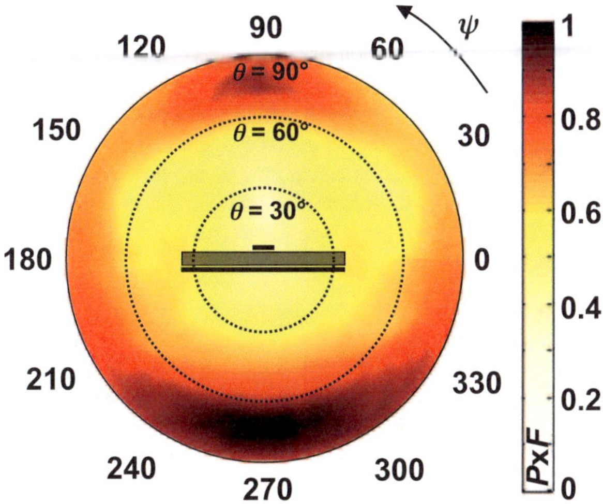

Figure 5.23: Normalized result for the performance joint criterion.

due to the antenna angular dependence, it is necessary to have a measure for quantifying this angular dependence. Through the analysis of the fidelity properties of UWB antennas it has been possible to determine the distortion that the transmitted signal suffers in the different angular directions with respect to the main beam direction in the time domain due to the non-ideal antenna behavior. From the developed analysis the areas where the signal shows high fidelity can be determined. Together with the fidelity, also the peak of the antenna impulse response has been regarded, since the fidelity does not give any information on the antenna gain.

In order to identify the areas where the antenna has good time domain behavior (low distortion of the radiated pulse and high peak energy) a joint criterion regarding both the fidelity and the pulse peak power has been introduced.

The developed analysis has been applied to two antennas (Vivaldi and Bow-tie) and measurements have been performed on a sphere. The Vivaldi antenna has a different behavior compared to the Bow-tie antenna with respect to the area with high performance. The pulse transmitted by the Vivaldi antenna shows one single focus area. For the Bow-tie antenna the area of high performance is the entire azimuth plane.

The performed analysis is important for communications and Radar as well. If the angular coverage of the system to be developed is known, suitable antennas can be chosen and be put in suitable orientations. Having the possibility to quantify the angular distortion also permits to introduce "correction" strategies at the receiver side.

6 Signal Optimization

In order to permit the coexistence of UWB systems with other systems, very low power transmission ($P_{\mathrm{max}} = -41.3$ dBm/MHz) is permitted for commercial devices. Hence, because of the power restriction given by the UWB mask (see Fig. 6.1), which limits the power spectral density instead of the total amount of transmit power, an efficient utilization of the available bandwidth is compulsory, i.e. the signal spectrum has to optimally exploit the allowed power spectral density in the transmit band in order to maximize the received SNR.

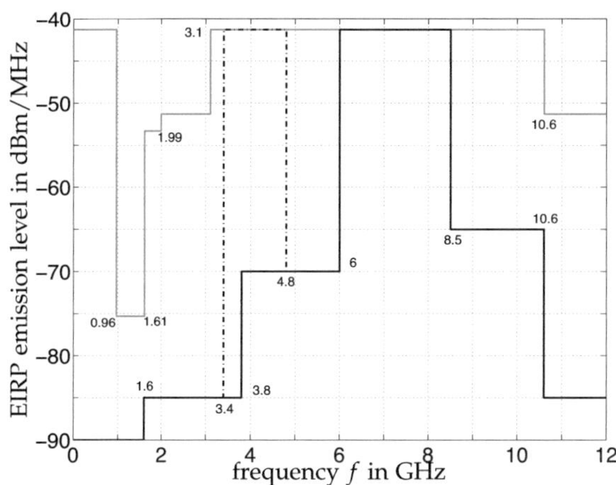

Figure 6.1: The UWB mask: FCC regulation (gray) and EU regulation without (black, solid line) and with low duty cycle (black, dotted line).

For impulse radio transmission a pulse shape that complies with the provided spectrum mask must be used. Typically, in UWB communication systems, the pulse from pulse generator has high spectral components at low frequencies. This means that its power spectral density cannot be assumed constant within the transmit frequency band. In addition, the radiation by the antenna is frequency dependent, which causes the transmitted signal to be further distorted, so that it may fit even less the given spectral mask. Hence, a compensation for the non-ideal behavior of the transmitter has to used.

In literature there are different methods to perform this compensation. A technique used to achieve this goal consists in digitally pre-distorting the pulse. This is performed by the utilization of digital filters [32], [70] - [72] creating an optimum signal through DSP implementation of digital FIR filters [71], which require immense computational effort. A second approach consists in directly designing a signal that exactly fulfills the UWB mask [33], [34]. Other methods are based on applying an analog filter to restrict the spectrum in the desired passband, having the filter a bandwidth (S_{21} parameter) that exactly matches the UWB mask [17] - [20]. In that case the spectrum is restricted by eliminating the undesired signal components outside the UWB range. However, through all these methods, the spectral power distribution inside the UWB mask is not improved. Nevertheless, the majority of these methods does not take into account the non-idealities of the UWB components at the transmitter side, in particular the antenna, which has a gain dependent on frequency, as seen in the previous chapter. Hence these methods are suboptimal.

In the following an UWB system based method, by which the transmit signal fits optimal to the given mask, is presented [31]. This is an advanced pulse shaping concept, that at the same time permits to compensate the non-ideal component behavior regarding the transfer function and hence to make the signal better exploit the maximum power of the UWB mask. A particular shaping filter is designed to be inserted in between the pulse generator and the antenna and its characteristic is calculated to compensate for the non-ideal behavior and distortions of the pulse generator and of the frequency-dependent antenna gain.

The chapter is organized as it follows. Firstly, the non-idealities of the RF-front end are regarded. Secondly, the advanced pulse shaping concept is mathematically described. Finally, it is verified through the development and the fabrication of practical pulse shaping filters for specific scenarios (pulse generator and two different antennas). The optimization of the power spectral desity of the radiated signal is verified through measurements.

6.1 Non-Idealities in the UWB RF-Front End

In the following, the non-ideal behavior of the UWB components at the transmitter side is analyzed.

Let consider the transmitter as schematically illustrated in Fig. 6.2 (top). It is composed of a signal generator and a UWB antenna. Let $p(t)$ be the pulse at the output of the pulse generator and $\mathbf{h}_A(t)$ the impulse response of the antenna. From what has been seen in the previous chapter and from [46], the effective on-air transmitted electric field is given by (ref. to eq. (5.1))

$$\mathbf{e}(t,r) = a(t,r) * \mathbf{h}_A(t) * \frac{1}{2\pi c_0} \frac{\partial}{\partial t} p(t) \tag{6.1}$$

where the derivative $\partial/\partial t$ is caused by the antenna. $a(t,r)$ is an attenuation term, which attenuates the wave and delays it, according to the distance r. The attenuation depends only on the distance r from the transmitter if only one specific direction of radiation is regarded. It coincides with the term defined in eq. (5.2), i.e. it produces an attenuation inversely proportional to the distance r and a delay proportional to r/c_0.

The previous relationship can be rewritten in the frequency domain as (ref. to eq. (5.6))

$$\mathbf{E}(f,r) = A(f,r) \cdot \mathbf{H}_A(f) \cdot \frac{1}{2\pi c_0} j\omega P(f) \tag{6.2}$$

where $P(f)$ is the spectrum of $p(t)$, $\mathbf{H}_A(f)$ is the antenna transfer function and the term $j\omega = j2\pi f$ results from the differentiation in the time domain. The term $A(f,r)$, which is the Fourier transform of $a(t,r)$, produces in the frequency domain an attenuation of entity $1/r$ and a phase shift of r/c_0.

The radiated wave must show a constant power spectral density in the licensed frequency range. However, even in the ideal case of an antenna with a frequency-independent transfer function in the UWB frequency range, the on air transmitted wave is given by the derivative of the pulse from the pulse generator.

Furthermore, the presence of a non-ideal antenna with a frequency dependent transfer function in the UWB passband causes the signal to be further distorted and to less fit the assigned mask. It must be also assumed that the pulse generator does not provide a signal with constant power spectral density. Fig. 6.3 exemplifies this concept. Here the spectrum $P(f)$ of the generated pulse is plotted (left) together with the non-ideal antenna transfer function (center). The generated pulse decreases with the frequency, since it is not possible to generate an ideally short pulse with an infinitely large bandwidth, while the antenna transfer function is not constant in the UWB frequency range, i.e. it is frequency dependent. The resulting on-air transmitted wave is plotted on the right side of Fig. 6.3, together with the required mask. Due to the non-ideal behavior of the components, the spectrum of the on-air transmitted wave does not optimally fit the given mask.

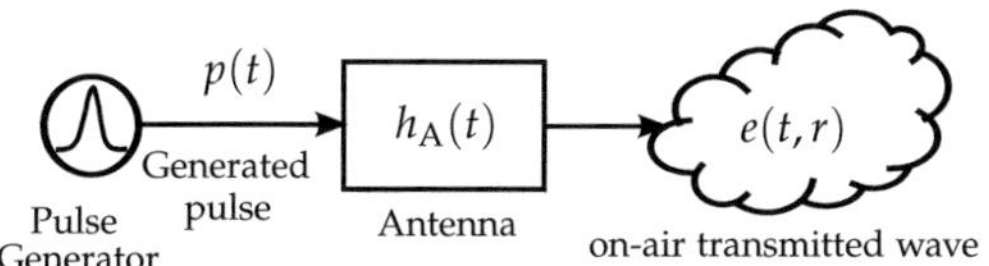

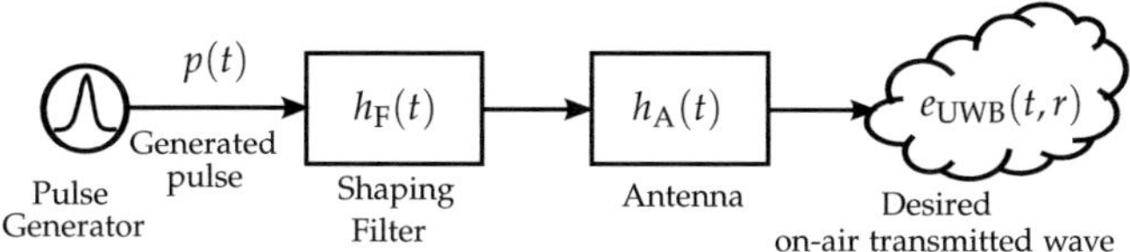

Figure 6.2: Schematic representation of the transmitter: standard realization (top), with insertion of shaping filter (bottom).

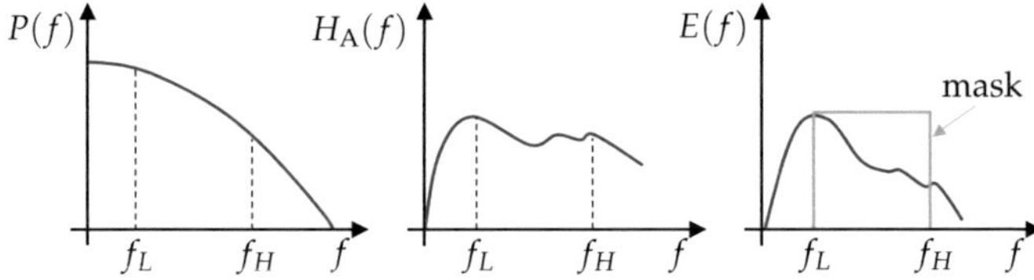

Figure 6.3: Effect of the non-ideal behavior of the transmitter: the spectrum $P(f)$ of the generated pulse (left), the antenna transfer function H_A (center) and the resulting normalized on-air transmitted signal $E(f)$ (right, see eq. (6.3)) together with the required mask.

6.2 Required Filter Transfer Function

The aim is that the on-air transmitted signal maximally fulfills the UWB mask. This can be achieved with a compensation of the antenna's and pulse generator's non-idealities.

It has to be observed that, in the filter design, the antenna transfer function in the direction of maximum gain (co-polarzation component) has to be considered because the limit for the radiated power has to be respected in each direction. Hence in the following the quantities $\mathbf{E}$ and $\mathbf{H}_A$ coincides with the main radiation direction component and are considered scalar. Moreover, $E(f,r)$ is normalized so that the dependence on the distance can be neglected, i.e. only relative values of power spectral density (PSD) are considered, which are normalized to the distance r, namely

$$E(f) = \frac{E(f,r)}{A(f,r)} \,. \tag{6.3}$$

This is done because the regulations impose that the transmitted signal must fulfill the given mask at the antenna output. Since the term $a(t,r)$ $(A(f,r))$ produces only a delay in time (phase shift in frequency) and an attenuation, and since the PSD is not influenced by phase shifts, the newly defined $E(f)$ is used in the developed analysis.

Let $s_{\mathrm{UWB}}(t)$ be the desired optimum on-air transmitted wave, i.e. the wave that perfectly fits the mask compensating also the components' non-idealities.

The optimum desired on-air transmitted wave can be achieved at the antenna output by introducing at the transmitter side a shaping filter with impulse response $h_F(t)$ (see Fig. 6.2, bottom) such that

$$s_{\mathrm{UWB}}(t) = h_A(t) * h_F(t) * \frac{1}{2\pi c_0}\frac{\partial}{\partial t}p(t) \,. \tag{6.4}$$

Rewriting the previous equation in the frequency domain, it gives

$$S_{\mathrm{UWB}}(f) = H_A(f) \cdot H_F(f) \cdot \frac{1}{2\pi c_0}j\omega P(f) \tag{6.5}$$

where $H_F(f)$ is the filter transfer function. Hence, it can be recognized that the filter transfer function that permits to obtain the desired on-air transmitted signal $s_{\mathrm{UWB}}(t)$

is given by

$$H_{\mathrm{F}}(f) = \frac{S_{\mathrm{UWB}}(f)}{H_{\mathrm{A}}(f) \cdot \frac{1}{2\pi c_0} j\omega P(f)} \, . \qquad (6.6)$$

This procedure compensates the effect of the non-ideal antenna behavior and takes into account the power spectral density of the signal generated by the pulse generator.

It has to be noticed that, comparing eq. (6.6) with eq. (6.2), the denominator of eq. (6.6) is exactly the transmitted signal in the case where no filter is applied, i.e.

$$H_{\mathrm{F}}(f) = \frac{S_{\mathrm{UWB}}(f)}{E(f)} \qquad (6.7)$$

where $E(f)$ is defined in eq. (6.3).

It has to be pointed out that in the ideal case of a constant antenna transfer function and a generated pulse whose derivative has constant power in the selected frequency interval, the required filter mask $H_{\mathrm{F}}(f)$ is simply a bandpass filter. On the other hand, due to the non-constant antenna transfer function and the non whiteness of the derivative of the generated pulse, the filter has to compensate for these non-idealities and hence it does not only have to select the desired band but also to pre-distort the signal. Hence, it is a "shaping filter" whose bandpass transfer function (S_{21}) is not constant.

6.3 Filter Design

In this section the application of the previous method for two different antennas is described. The presented analysis is applied for the EU UWB case.

The considered pulse $p(t)$ is the one created by a commercially available pulse generator (PSPL 3600) [80]. In Fig. 6.4 it is plotted together with its normalized spectrum.

Two different antennas have been taken into account, which have different radiation principles: a travelling wave antenna (Vivaldi antenna) and a small antenna (Bow-tie antenna). As it has been seen in the previous chapter, these antennas radiate different signals in the different angular directions. In the filter design the transfer function in the direction of maximum gain (co-polarzation component) has to be considered because the limit for the radiated power has to be respected in each direction.

6.3.1 Vivaldi Antenna

The measured Vivaldi antenna transfer function[1] is shown in Fig. 6.5(a).

In order to calculate the required filter transfer function, eq. (6.7) has been used. Instead of multiplying the various terms H_{A}, $j\omega$ and $P(f)$, since the denominator of eq. (6.6) is exactly the transmitted signal in the case when no filter is applied, the signal $E(f)$ transmitted by the system composed of the pulse generator and the Vivaldi antenna (Fig. 6.2 top) has been directly measured. The measured $E(f)$ is illustrated in Fig. 6.5(b) (solid line), restricted to the frequency interval of interest from

---

[1]It has been obtained as described in chapter 4.

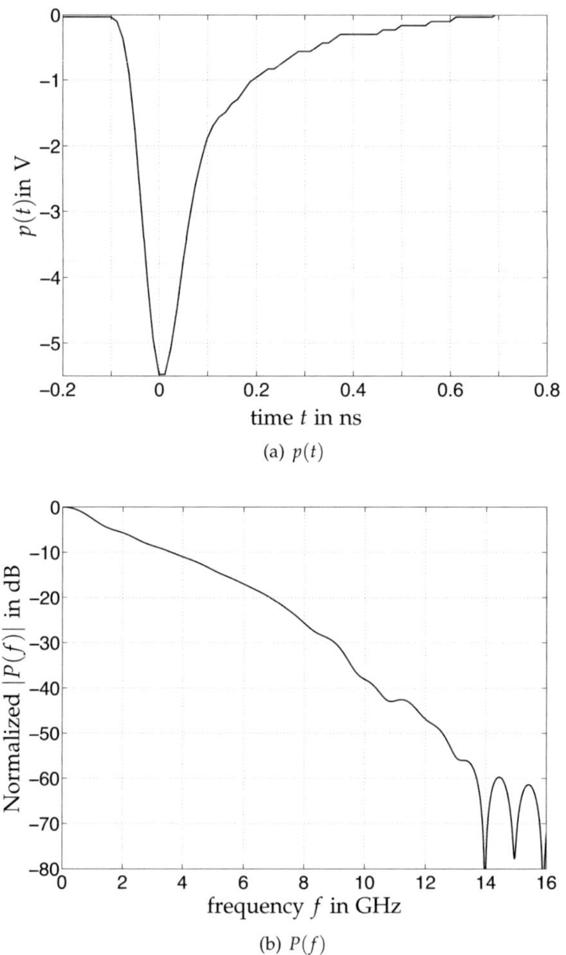

(a) $p(t)$

(b) $P(f)$

Figure 6.4: The pulse created by the pulse generator: time domain pulse shape $p(t)$ and frequency domain normalized spectrum $|P(f)|$.

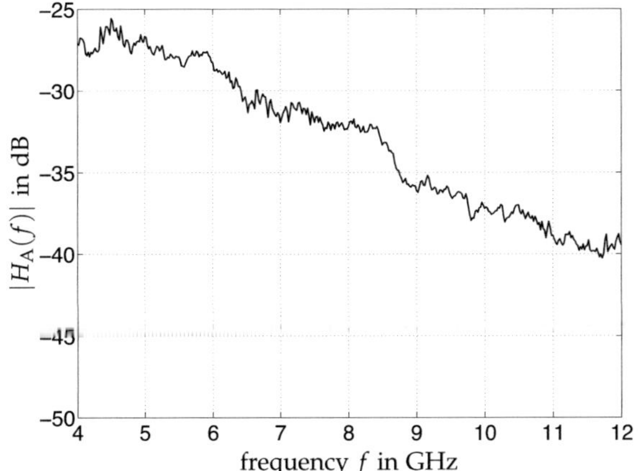

(a) Measured Vivaldi antenna transfer function (main beam direction, co-polarization)

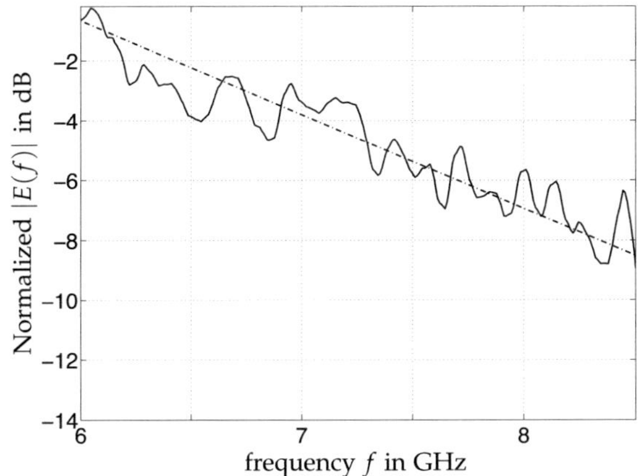

(b) Measured normalized transmitted spectrum $|E(f)|$ (solid line) and its linear approximation (dotted line) for main beam direction, co-polarization

Figure 6.5: Vivaldi antenna: transfer function in the main beam direction, co-polarized component (a) and normalized transmitted spectrum $|E(f)|$, given by eq. (6.2), when no filter is applied ((b), solid line) together with its linear approximation ((b), dotted line).

6 to 8.5 GHz and normalized in that band. For the design of the filter, this spectrum has then been approximated by a linear interpolation, as shown in Fig. 6.5(b) (dotted line). The radiated power spectral density is decreasing by around 8 dB over the 2.5 GHz bandwidth. This interpolated curve has been inserted in eq. (6.6). As goal function, $S_{UWB}(f)$ has been considered having a rectangular shape in the $6-8.5$ GHz frequency interval. The obtained required filter mask is plotted in Fig. 6.6. A particular shaping filter is constructed for the Vivaldi antenna and verified through measurements, as it is described in the following.

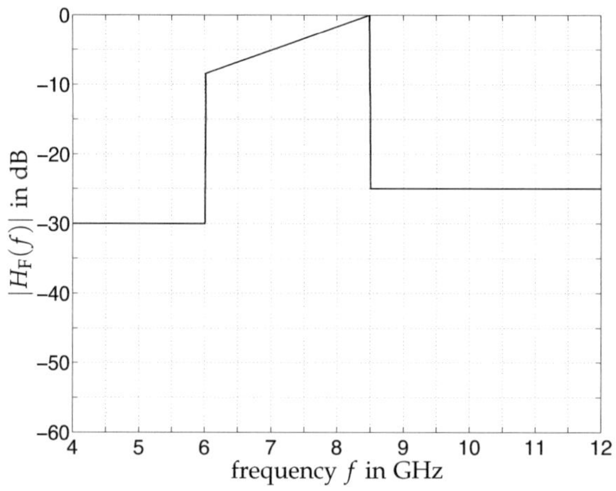

Figure 6.6: Mask of the required shaping filter for the Vivaldi antenna.

6.3.2 Bow-tie Antenna

The Bow-tie antenna presents a transfer function (see Fig. 6.7(a)) with higher decreasing slope with respect to the Vivaldi antenna in the considered frequency interval. The effective electric field $E(f)$ radiated by the system composed by the pulse generator and the Bow-tie antenna has the shape given in Fig. 6.7(b). For this antenna the decrease of the radiated power spectral density amounts to 14 dB over 2.5 GHz of bandwidth. Hence, considering its linear approximation (dotted line in Fig. 6.7(b)), it can be seen that the filter to be developed has a shape which is sharper compared to the one required by the Vivaldi antenna. Also in this case, as goal function $S_{UWB}(f)$ has been considered having a rectangular shape in the $6-8.5$ GHz frequency interval. Applying eq. (6.7) with the measured $E(f)$ and the so defined desired goal function $S_{UWB}(f)$, the resulting required filter shape is shown in Fig. 6.8. In the following, the description of a realized filter for the Bow-tie antenna is given, together with the verification through measurement results.

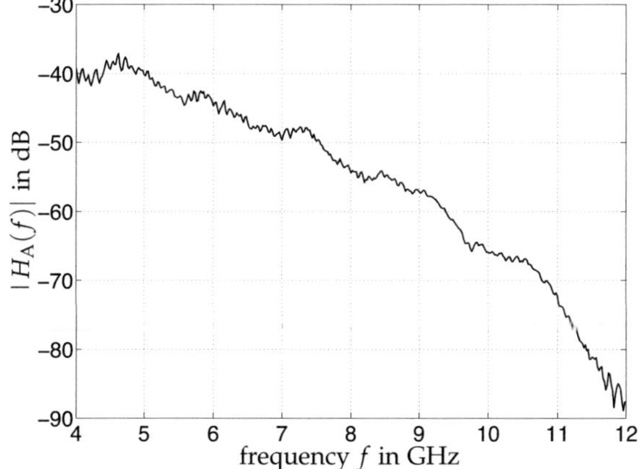

(a) Measured Bow-tie antenna transfer function (main beam direction, co-polarization)

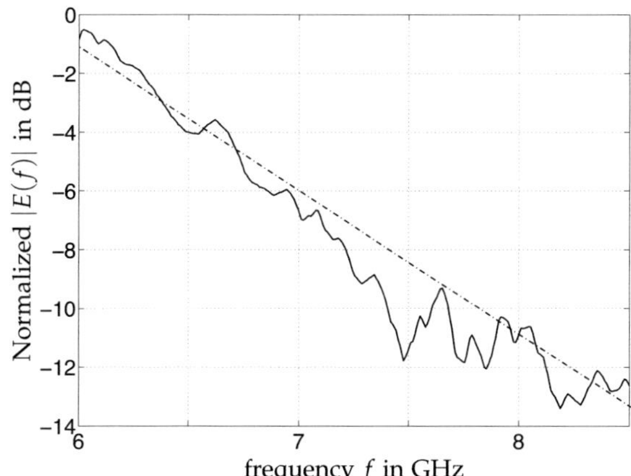

(b) Measured normalized transmitted spectrum $|E(f)|$ (solid line) and its linear approximation (dotted line) for main beam direction, co-polarization

Figure 6.7: Bow-tie antenna: transfer function in the main beam direction, co-polarized component (a) and normalized transmitted spectrum $|E(f)|$, given by eq. (6.2), when no filter is applied ((b), solid line) together with its linear approximation ((b), dotted line).

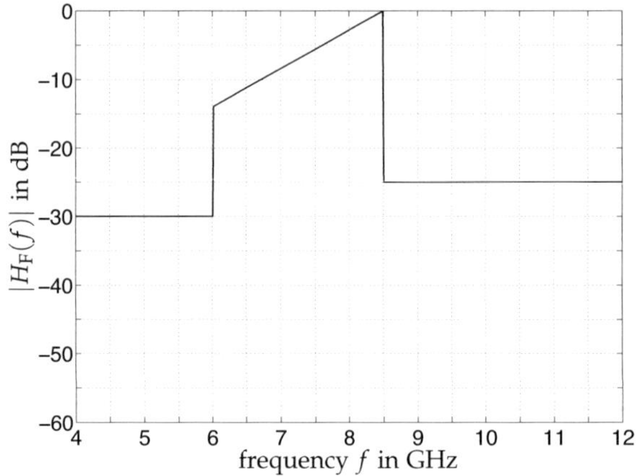

Figure 6.8: Mask of the required shaping filter for the Bow-tie antenna.

6.3.3 Realized Filters

6.3.3.1 Shaping Filter for the Vivaldi Antenna

The realized shaping filter for the Vivaldi antenna is composed by the cascade of a bandpass section and a highpass section, which are now in detail regarded.

Bandpass Section The bandpass section permits to select the desired frequency interval. The selected structure used to implement this section is the bandpass filter presented in chapter 3 (ref. to section 3.5.3, Model 2). The highpass section has ben introduced in order to obtain the desired "shaping". The choice of the structure described in chapter 3 has been done in order to have directly the possibility to compare the behavior of the shaping filter with the behavior of the bandpass filter already investigated in detail previously, and hence to quantify the impact of this particular filter structure with respect to the classical bandpass filter structure.

Highpass Section The highpass section has been constructed such that the cut-off frequency f_{cutoff} of the filter transmit band is exactly $f_{\text{H}} = 8.5$ GHz, the upper frequency of the EU UWB interval. It has been realized through short circuited stubs. The length of these stubs has been selected to be quarter wavelength at the frequency $f_{\text{cutoff}} = f_{\text{H}} = 8.5$ GHz. The number n of stubs has been choosen in order to obtain the desired in-band slope of 8.5 dB of the filter mask. After computer simulations, it has been seen that the optimum stub number $n = 3$. Higher n produces a more increasing slope while lower n a too slowly increasing slope. The stubs are spaced with connecting lines of length $\lambda_{\text{cutoff}}/4$.

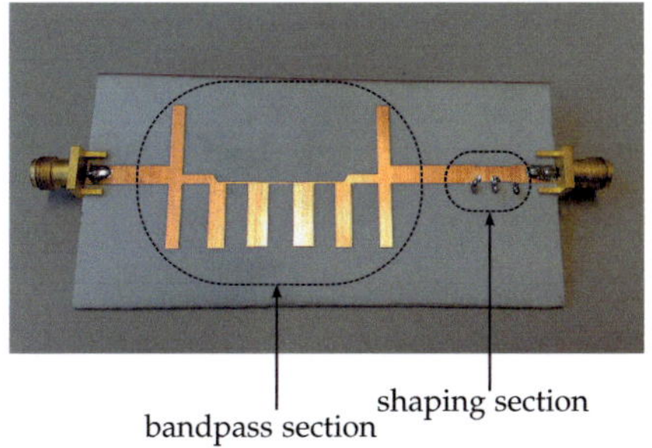

Figure 6.9: Fabricated shaping filter for the Vivaldi antenna.

One prototype of the developed shaping filter has been fabricated on a Rogers substrate RO4003 with $\varepsilon_r = 3.38$ and thickness 1.57 mm and it is shown in Fig. 6.9. Measurements on the realized filter has been performed to validate its behavior. In Fig. 6.10 the absolute value of the measured S_{21} parameter is plotted. As it can be recognized, it satisfies the derived theoretical mask.

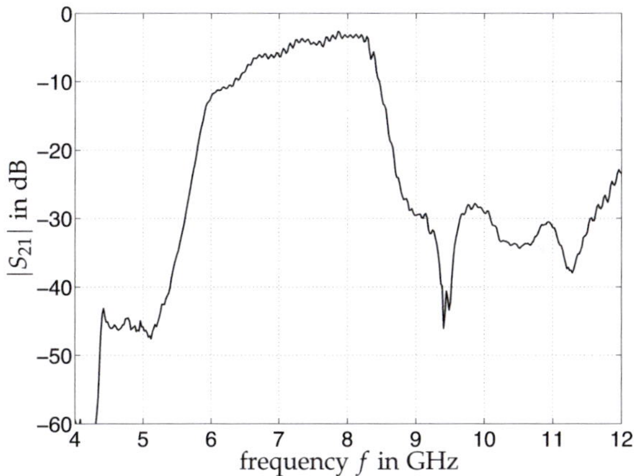

Figure 6.10: Measured $|S_{21}|$ parameter of the fabricated shaping filter for the Vivaldi antenna.

In Fig. 6.11 the GDT of the realized shaping filter, obtained according to eq. (2.3), is plotted.

6.3.3.2 Shaping Filter for the Bow-tie Antenna

Differently from the Vivaldi antenna, in the case of the Bow-tie antenna the required shaping filter needs a slop of 14 dB in the EU UWB band. The shaping filter realized for

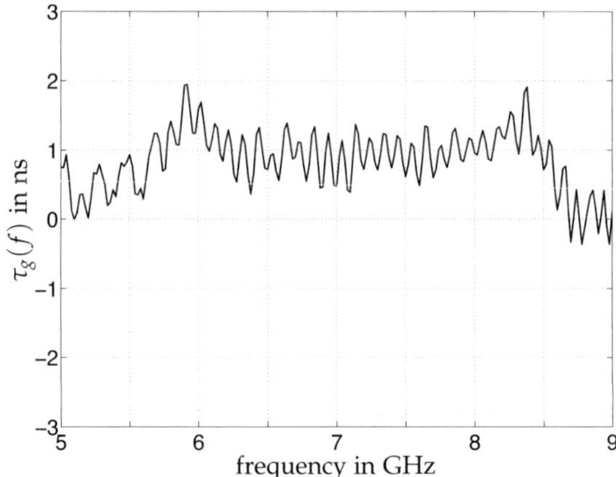

Figure 6.11: Measured GDT $\tau_g(f)$ of the fabricated shaping filter for the Vivaldi antenna.

the Bow-tie antenna consists in a CPW structure, which is composed by the integration of a highpass section and a lowpass section. These sections have been constructed starting from the base CPW discontinuities presented in chapter 3 (ref. to section 3.5.5) and are now in detailed regarded.

Highpass Section This section consists in the integration of two different CPW elements with highpass behavior: the open-end series stub and the short-end shunt stub. These elements are integrated one into the other for space reduction. The lengths of the stubs have been optimized in order to obtain the desired slope in the passband.

Lowpass Section This section has been realized through a short-end series stub and a defected ground structure. The defected ground structure has been realized through slots etched symmetrically in the CPW ground plane. The slots have been bended to save space. This basic structure has been repeated twice to obtain a sharper upper transitional band.

It has to be pointed out that the realization of the shaping filter for the Bow-tie antenna differs in principle from the adopted realization in the case of the Vivaldi antenna. For the Vivaldi antenna the desired filter shape has been obtained through the cascade of a bandpass section, for selecting the required EU UWB band, and a highpass section, for obtaining the desired slope of the in-band filter mask. Hence, this additional highpass section deforms the constant mask of the bandpass filter. In the case of the Bow-tie antenna the shaping operation and the band-selection operation are performed at the same stage and are hence integrated.

A prototype of this filter has been etched on a substrate Rogers RO4003, $\varepsilon_r = 3.38$, thickness $h = 1.57$ mm (see Fig. 6.12). Its frequency domain behavior has been tested

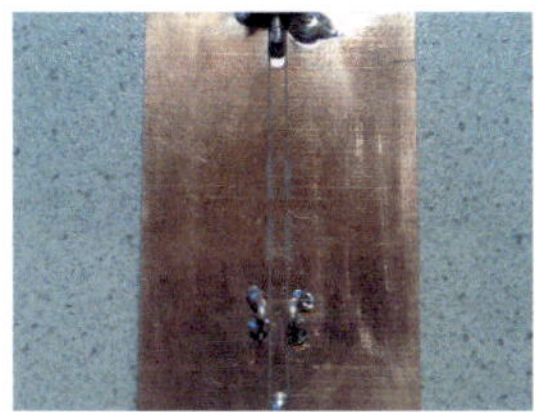

Figure 6.12: Fabricated shaping filter for the Bow-tie antenna.

through measurement. The absolute values of the measured S_{21} and S_{11} parameters are shown in Fig. 6.13. As desired, it satisfies the required mask, i.e. it has a slope of approximately 14 dB in the EU UWB mask. Moreover, it selects the EU UWB frequency interval. Its GDT is plotted in Fig. 6.14.

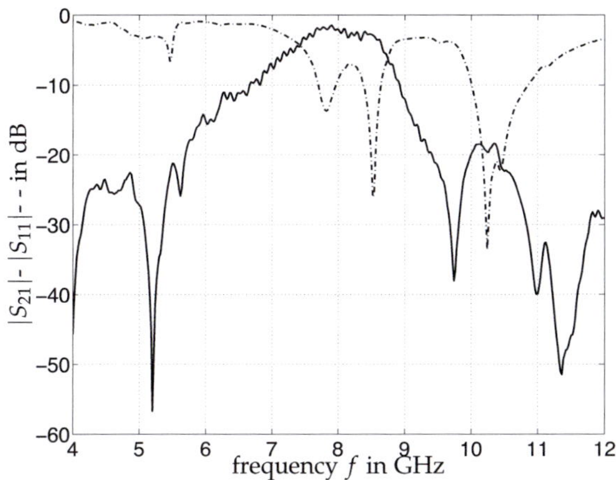

Figure 6.13: Measured $|S_{21}|$ (solid line) and $|S_{11}|$ (dotted line) parameters of the fabricated shaping filter for the Bow-tie antenna.

6.4 Verification

Since the developed filters are intended to optimize the spectrum for impulse radio transmissions with a particular pulse generator, their operability has to be evaluated by measuring the spectrum of the radiated waves.

In this section the measurement system configuration for the verification measurements is presented. With the evaluation of the operability of the filter prototypes also

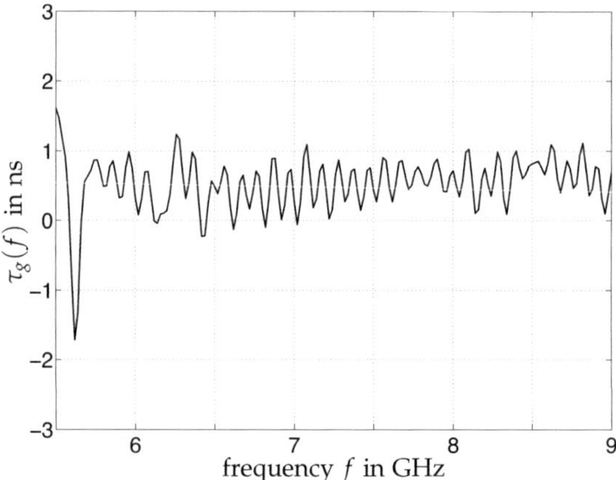

Figure 6.14: Measured GDT $\tau_g(f)$ of the fabricated shaping filter for the Bow-tie antenna.

the applicability of the entire pulse shaping filter concept is proven. In Fig. 6.15 the measurement scenario is illustrated.

The system setup is composed, at the transmiter side, by a Trigger generator (Tektronix, with 2 GHz Function/Arbitrary Waveform Generator) and a Pulse Generator (Pico Second Pulse Labs PSPL 3600). The pulse shape of this generator has been used in the previous synthesis of the filters (see Fig. 6.4). The last element at the transmitter side is the transmit antenna (positioned at a fixed position, orientation and distance to the receiver, for all the performed measurements). The transmit antenna is either the Bow-tie antenna or the Vivaldi antenna.

At the receiver side there is a receive antenna and an oscilloscope (Agilent Infinium DCA, 40 GSa/s, with a dynamic of about 60 dB (12 bit) and a bandwidth of 12 GHz). The oscilloscope is synchronized to the trigger signal which is fed to the oscilloscope over a cable connection from the trigger generator. The distance between the transmit and the receive antenna is $R = 2.9$ m.

In the performed measurements a horn antenna (Model 6100) has been used at the receiver. This antenna has been selected due to its directivity (that permits to avoid spurious reflections from the surrounding) and its almost frequency-independent transfer function and good decoupling between the two polarizations, as it can be seen in Fig. 6.16. The transfer function of the horn antenna has been recovered through the two-antennas method, as described in chapter 4 (ref. to section 4.1.1.1).

The system is controlled via a laptop computer which also permits the data acquisition.

At the transmiter side three different configurations are possible

1. Direct Connection (A-1-1-B): In this configuration the transmit antenna is directly connected to the pulse generator. Consequently, it is possible to directly measure

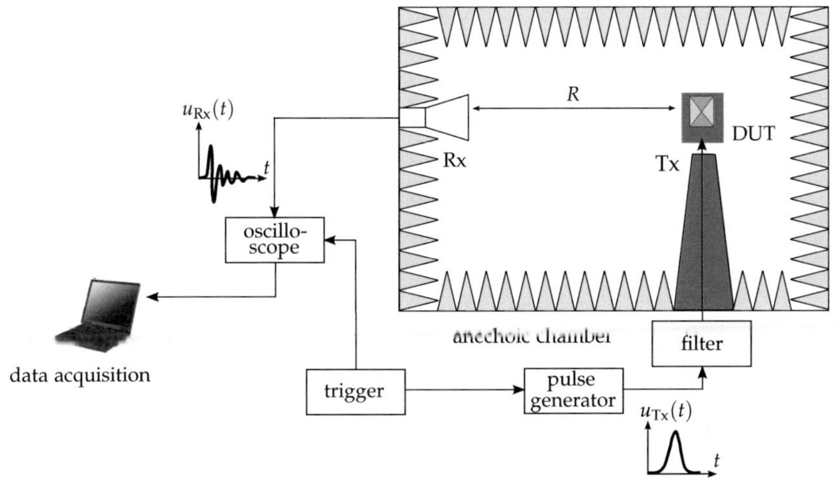

Figure 6.15: Measurement scenario for verification in the time domain.

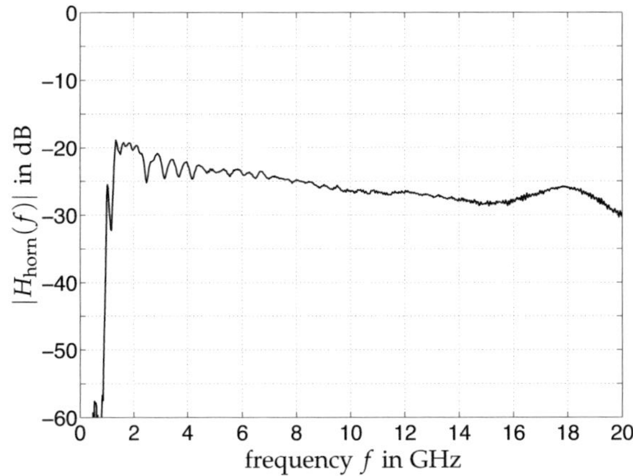

Figure 6.16: Measured transfer function of the used horn antenna (co-polarization).

the pulse spectrum $E(f)$ transmitted by the antenna.

2. Bandpass Filter (A-2-2-B): In this configuration, a bandpass filter is inserted between the pulse generator and the antenna. The cable connection has been made as short as possible, in order to minimize additional losses and attenuation in the cables. Furthermore, an absorber has been placed in front of the filter in order to avoid radiation from the filter. In this configuration it is possible to directly evaluate the impact of the bandpass filter on the system performance.

3. Shaping Filter (A-3-3-B): In this configuration a shaping filter is connected in between the pulse generator and the transmit antenna. The cable connection has been made as short as possible in order to minimize additional cable losses and attenuation. Moreover, the filter has been covered by absorbing material in order to prevent that radiation from the filter influences the measurements. In this configuration it is possible to directly quantify the influence of the shaping filter on the system performance.

6.4.1 Data Acquisition

The data have been collected in the time domain. The selected time interval has been between -15 ns and 15 ns relative to the trigger signal, with $N = 19.210$ samples at a sampling rate of 40 Gs/s. An average on 128 acquisitions has been made. The signal recovered by the oscilloscope, $u_{\text{Rx}}(t)$, according to the block scheme of Fig. 6.15 and to the UWB link description of the previous chapter (ref. to eq. (5.4)), is in the main beam direction (without filter)

$$u_{\text{Rx}}(t) = h_{\text{horn}}(t) * h_{\text{Ch}}(t) * h_{\text{AUT}}(t) * \frac{1}{2\pi c_0} \frac{\partial}{\partial t} p(t) \tag{6.8}$$

where $h_{\text{horn}}(t)$ is the impulse response of the used horn antenna, $h_{\text{Ch}}(t)$ is the channel impulse response (ref. eq. (5.2)), h_{AUT} is the impulse response of the Antenna Under Test (Vivaldi or Bow-tie) and $p(t)$ is the pulse from the pulse generator. If a filter (either a bandpass or a shaping filter) with impulse response $h_{\text{F}}(t)$ is inserted in between the pulse generator and the antenna, the previous equation has to be modified in the following way

$$u_{\text{Rx}}(t) = h_{\text{horn}}(t) * h_{\text{Ch}}(t) * h_{\text{AUT}}(t) * h_{\text{F}}(t) * \frac{1}{2\pi c_0} \frac{\partial}{\partial t} p(t) . \tag{6.9}$$

6.4.1.1 Elimination of the receive antenna influence

In oder to obtain the effective on-air transmitted signal, the influence of the receive horn antenna has to be eliminated. This is done by post-processing the acquired data. Firstly, the time domain acquired data are transformed into the frequency domain, using a Discrete Fourier Transform (DFT). Letting Δt be the time spacing of the acquired data and N the number of sampled points, the respective sampling frequency is $F_s = 1/\Delta t$ and the frequency resolution is $\Delta f = F_s/N \simeq 2$ MHz. In order to obtain the frequency domain spectrum $U_{\text{Rx}}(f)$, the DFT is implemented in the following way (ref. to eq. (A.3))

$$U_{\text{Rx}}(k\Delta f) = \frac{1}{\Delta t} \sum_{n=0}^{N-1} u_{\text{Rx}}(n\Delta t) \cdot \exp\left[-j\frac{2\pi}{N}kn\right] \qquad (6.10)$$

following the procedure described in section A.2. Then, in order to eliminate the antenna influence, the obtained frequency domain data has been divided by the horn antenna transfer function as

$$\tilde{U}_{\text{Rx}}(f) = \frac{U_{\text{Rx}}(f)}{H_{\text{horn}}(f)} \qquad (6.11)$$

for $f = k\Delta f$. The attenuation due to the distance $R = 2.9$ m between the transmit and the receive antenna has also been compensated. Moreover, the obtained signal has been interpolated so that the frequency resolution is $\Delta f = 1$ MHz.

6.4.2 Power Spectral Density

Once the mathematical background has been described, the obtained results for the PSD calculated from the time domain measurement data are now presented.

6.4.2.1 Vivaldi Antenna

Three measurements have been performed. Firstly, the antenna alone, in the (A-1-1-B) configuration, in order to obtain the time domain samples of the received signal resulting from a transmission without filter. This data has been used in the filter synthesis, as previously explained. Secondly, a bandpass filter has been inserted between the pulse generator and the antenna (configuration (A-2-2-B)). This has been done in order to compare the frequency domain and time domain impact of the shaping filter with respect to a classical bandpass filter. Finally, the realized shaping filter has been inserted (configuration (A-3-3-B)).

In Fig. 6.17(a) the obtained normalized received signal spectrum for the configuration (A-2-2-B) is plotted. The spectrum has been calculated as explained in the previous section. Since the aim of the filter is to shape the spectrum according to the UWB mask, the absolute value of the spectrum, which depends on the distance between the antennas, is not of practical interest here. The used bandpass filter is the one described in chapter 3 (ref. to section 3.5.3, Model 2). In Fig. 6.17(b) the obtained result for the configuration (A-3-3-B) is shown. Directly from this picture it can be recognized that the shaping filter permits a better exploitation of the in-band spectrum, i.e. the mask is better fulfilled.

In order to quantify the system performance improvement by using the shaping filter with respect to the classical bandpass filter, the efficiency of the system has been calculated. This has been done by integrating the power spectral density of the signal in the passband interval $6 - 8.5$ GHz and comparing the result to the integral of the power spectral density of an ideal signal having rectangular spectrum in the $6 - 8.5$ GHz interval and a maximum value equal to the maximum value of the spectrum of the measured signal in that band. The obtained results are summarized in Tab.

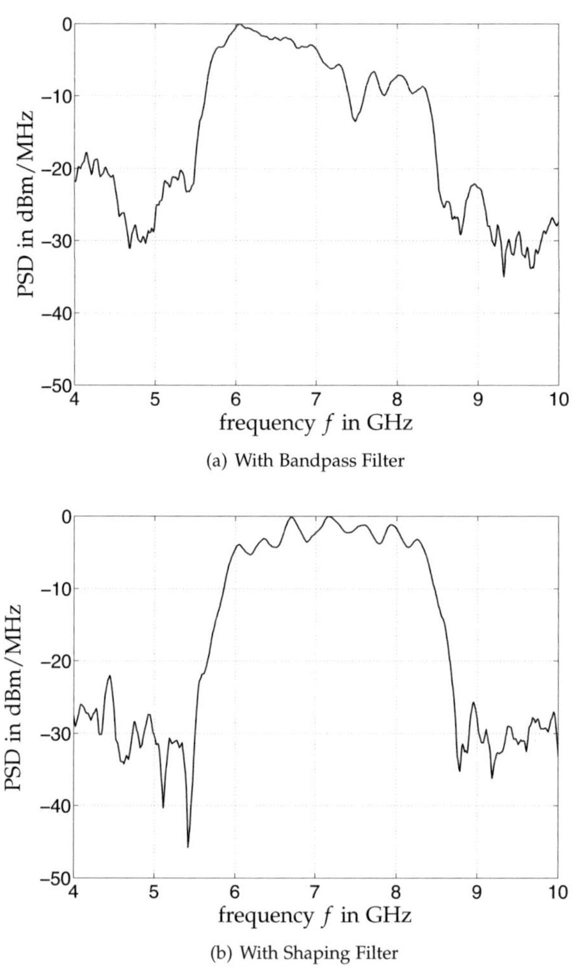

(a) With Bandpass Filter

(b) With Shaping Filter

Figure 6.17: Calculated normalized PSD of the radiated wave: (a) with bandpass filter (configuration (A-2-2-B)) and (b) with shaping filter (configuration (A-3-3-B)) for the Vivaldi antenna.

6.1. The case "Without-Filter" refers to the measurement in configuration (A-1-1-B). In that case, in order to respect the mask, the power of the transmitted signal has to be lowered.

Table 6.1: Vivaldi antenna: computed efficiency of the system in the three different configurations.

Without Filter	Bandpass Filter	Shaping Filter
1.6 %	35.5%	66.4%

From the obtained results, it can be concluded that the application of the shaping filter permits to obtain a considerable improvement of the system performance, since the mask is better fulfilled with respect to the case when only a bandpass filter is applied.

6.4.2.2 Bow-tie Antenna

Also in the case of the Bow-tie antenna, three measurements have been performed. Firstly, the antenna alone (configuration (A-1-1-B)). Secondly, a bandpass filter (the one described in section 3.5.5.2) has been added (configuration (A-2-2-B)). Finally, with the realized shaping filter (configuration (A-3-3-B)). The obtained results are plotted in Fig. 6.18.

In order to evaluate the improvement of the usage of the shaping filter with respect to the classical bandpass filter, the efficiency of the system has been calculated as previously described. The obtained results are summarized in Tab. 6.2.

Table 6.2: Bow-tie antenna: Computed efficiency of the system in the three different configurations.

Without Filter	Bandpass Filter	Shaping Filter
1.6 %	38.2%	58.2%

From these results, it can be concluded that the usage of the shaping filter permits the mask to be much better fulfilled than in the case when only a bandpass filter is applied and hence to obtain the improvement of the system performance.

6.4.3 Radiated Pulse Shape

Together with the frequency domain performance, also the time domain behavior of the application of the shaping filter is investigated in order to quantify the additional distortion of the transmitted pulse caused by the shaping filter.

To obtain the time domain behavior of the proposed shaping filter and to quantify its influence with respect to the case where no filter is applied, starting from the measured

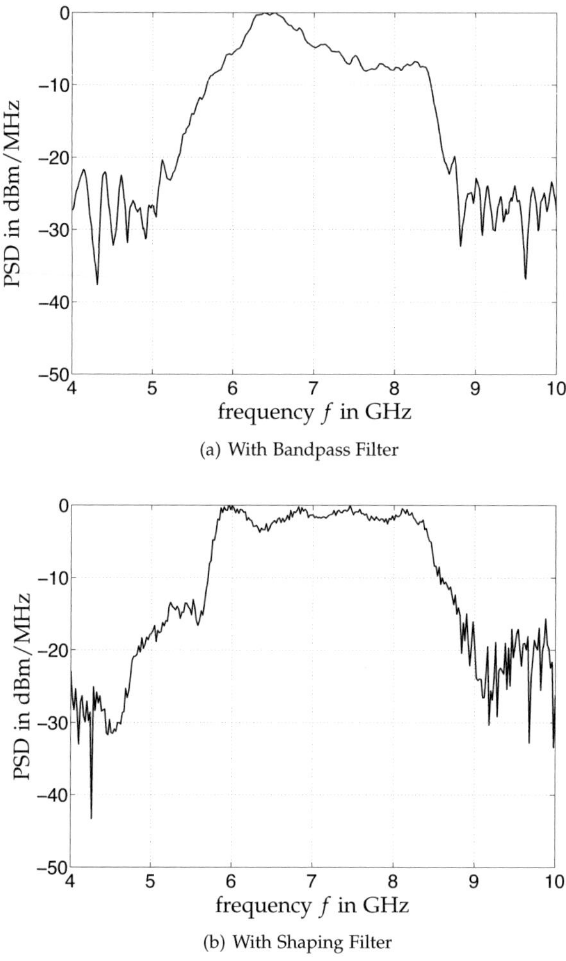

(a) With Bandpass Filter

(b) With Shaping Filter

Figure 6.18: Calculated normalized PSD of the radiated wave: (a) with bandpass filter (configuration (A-2-2-B)) and (b) with shaping filter (configuration (A-3-3-B)) for the Bow-tie antenna.

signal $u_{Rx}(t)$ it is necessary to eliminate the influence of the receive horn antenna and of the channel. Since in eq. (6.11) the horn antenna and the channel influence has already been eliminated, starting from this equation, the respective time domain wave $\tilde{u}_{Rx}$ has been recovered. The obtained $\tilde{U}_{Rx}$ from eq. (6.11) has been previously limited to the $2 - 18$ GHz frequency interval[2]. Now, for the transformation back to the time domain, a zero padding between $0 - 2$ GHz is inserted and an additional zero padding of ten times the length of the signal is attached after the samples, in order to increase the time domain resolution. To the obtained wave $\tilde{U}_{Rx,pad}$ the IDFT is applied (ref. to eq. (A.3)), namely

$$\tilde{u}_{Rx}(n\Delta t) = \frac{1}{\Delta f} \sum_{k=0}^{N-1} \tilde{U}_{Rx,pad}(k\Delta f) \cdot \exp\left[j\frac{2\pi}{N}nk\right] . \tag{6.12}$$

Once the on-air transmitted wave in the time domain has been recovered, the time domain parameters defined in chapter 2 (peak P, FWHM τ_{FWHM} and ringing $\tau_{0.1}$) have been calculated for the system in configuration (A-1-1-B), i.e. without filter, in configuration (A-2-2-B), i.e. with bandpass filter, and in configuration (A-3-3-B), i.e. with shaping filter, in oder to quantify the impact of the shaping filter in the time domain.

6.4.3.1 Vivaldi Antenna

Following the procedure described previously, the obtained on-air transmitted waves without filter, with bandpass filter and with shaping filter are plotted in Fig. 6.19. The calculated time domain parameters are summarized in Tab. 6.3.

Table 6.3: Vivaldi antenna: calculated time domain parameters of the on-air transmitted wave.

	Peak P in V	τ_{FWHM} in ps	$\tau_{0.1}$ in ns
Without Filter	0.0949	160	0.45
Bandpass Filter	0.0187	490	2.98
Shaping Filter	0.0164	250	2.78

As it can be evinced, the application of a filter (either a bandpass or a shaping filter) causes the decrement of the peak. This is due to the fact that the filter, selecting a particular frequency interval $f_L < f < f_H$, limits the transmitted power, since the original transmitted signal has power content also in the intervals $f < f_L$ and $f > f_H$ in particular at lower frequencies. Moreover, the shaping filter causes a slightly higher decrement of the peak with respect to the bandpass filter (the peak P with shaping filter is approximatively 10% lower than in the case of the bandpass filter). This is due to the fact that the shaping filter has a non flat S_{21} in the bandpass $f_L < f < f_H$ interval and this causes a further attenuation. However, the increment of the FWHM and of

---

[2]Frequencies below 2 GHz cannot be regarded because the used horn antenna does not work below this limit.

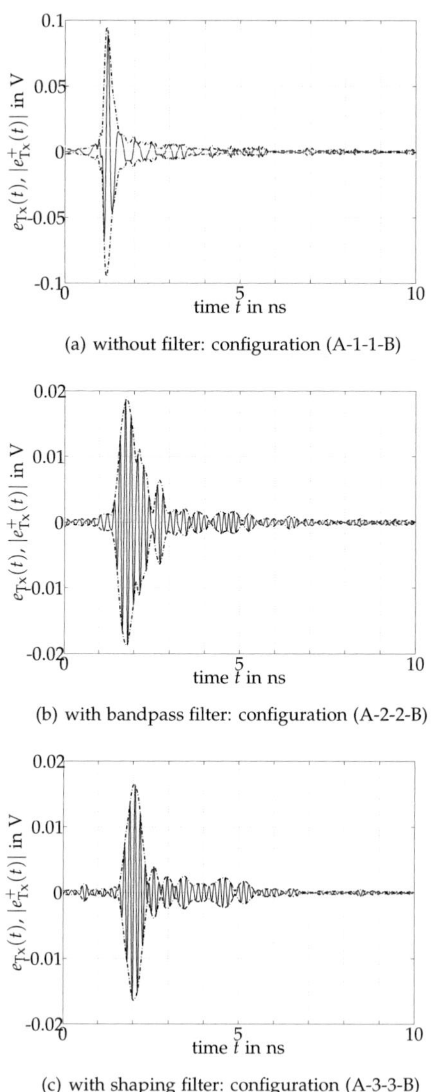

(a) without filter: configuration (A-1-1-B)

(b) with bandpass filter: configuration (A-2-2-B)

(c) with shaping filter: configuration (A-3-3-B)

Figure 6.19: Measured on-air transmitted wave (solid line) and its envelope (dotted line) in the three configurations for the Vivaldi antenna at the antenna output.

the ringing in the case of the shaping filter is smaller than in the case of the bandpass filter.

6.4.3.2 Bow-tie Antenna

Analogously to the Vivaldi antenna, following the procedure previously described, the obtained on-air transmitted waves for the Bow-tie antenna without filter, with bandpass filter and with shaping filter are plotted in Fig. 6.20.

The time domain parameters have been calculated also in the case of the Bow-tie antenna and the obtained results are summarized in Tab. 6.4

Table 6.4: Bow-tie antenna: calculated time domain parameters of the on-air transmitted signal.

	Peak P in V	τ_{FWHM} in ps	$\tau_{0.1}$ in ns
Without Filter	0.0533	170	0.41
Bandpass Filter	0.0115	260	1.24
Shaping Filter	0.0114	270	1.96

Also in this case, the application of a filter (either a bandpass or a shaping filter) causes a decrement of the peak, as previously described. The shaping filter causes a slightly higher reduction of the peak with respect to the bandpass filter. The FWHM and ringing increase with respect to the case of the configuration (A-1-1-B), where no filter is applied. However, these parameters are slightly higher for the shaping filter with respect to the bandpass filter.

However, in the time domain the performance is slightly lower with respect to the bandpass filter since the applied shaping filter requires a signal attenuation in order to obtain the desired frequency domain shape.

6.5 Conclusion

In this chapter an UWB system based method is presented, by which the transmit signal fits optimally to the given mask. This advanced pulse shaping concept permits to simultaneously compensate for the non-ideal behavior of the antenna and the pulse generator and to fit the signal better to the given frequency mask. A complete mathematical description of the method has been given. This method consists in realizing a filter, which not only selects the frequency interval of interest (as a classical bandpass filter) but also modifies the pulse shape in order to obtain a better fulfillment of the given mask in the frequency domain. This method has been applied to two particular antennas and *ad hoc* shaping filters have been fabricated. The method has been validated by measurement results. The developed filters effectively improve the spectral characteristics of the transmitted pulse regarding the given UWB mask. In particular the non-ideal characteristics of the transmitter (pulse generator and antenna) can

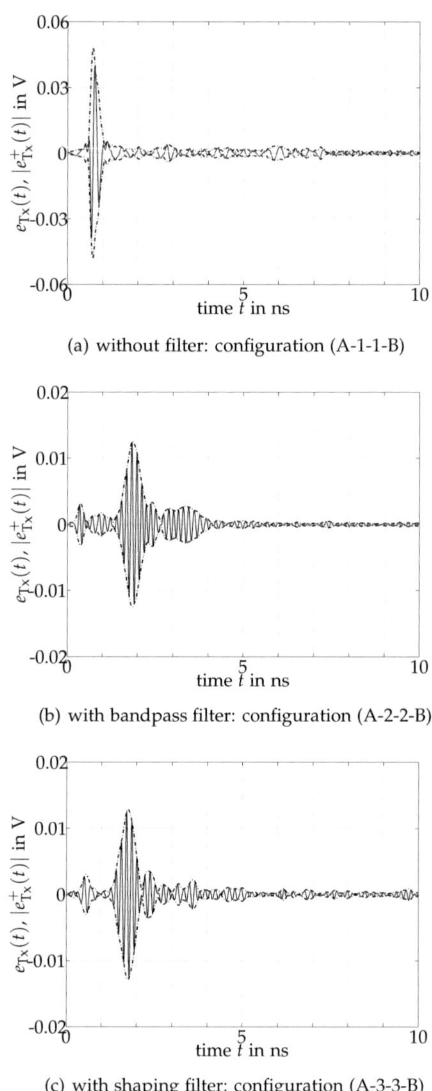

(a) without filter: configuration (A-1-1-B)

(b) with bandpass filter: configuration (A-2-2-B)

(c) with shaping filter: configuration (A-3-3-B)

Figure 6.20: Measured on-air transmitted wave (solid line) and its envelope (dotted line) in the three configurations for the Bow-tie antenna at the antenna output.

be effectively compensated with the developed pulse shaping filter structures. By regarding the radiated signals in the time domain it has been shown that the additional pulse distortion caused by a shaping filter compared to the distortion introduced by a conventional bandpass filter is negligible.

7 UWB Time Domain Radar Calibration

During the last years UWB technology has received increasing interest not only for communications applications but also for Radar purposes. For Radar applications the extremely large bandwidth permits a high range resolution and gives together with polarimetry the possibility of target and object recognition and classification. The fully polarimetric Radar calibration in the frequency domain has been a research topic at the end of the eighties [73], [74]. For UWB time domain Radar calibration these results are not appropriate, but the basic ideas of these research activities stimulate the UWB time domain Radar calibration.

In the majority of cases, UWB Radar systems have been analyzed in the frequency domain. Early investigations in the time domain have been presented in [39], where a time domain Radar equation has been given, although not in a polarimetric form. In [42] values of Radar cross sections obtained from time domain measurements are presented. The author of [77] published several articles on time domain Radar, but did not consider a polarimetric calibration. In summary the most important information contributions to UWB time domain Radar target characteristics have been lost.

No fully polarimetric time domain UWB Radar calibration has been found in the literature before the publication of [43]. The typical frequency domain characterization is only valid for a single frequency or a relatively narrow bandwidth and cannot directly be extended to UWB impulse Radar calibration. For UWB Radar a new calibration approach in the time domain is required, in order that UWB time domain Radar measurements and target classifications can be performed directly through time domain measurements with impulse Radar systems. It has to be pointed out that performing UWB Radar operations, both calibration and measurements, directly in the time domain has several advantages with respect to classical frequency domain measurements. Polarimetric impulse Radar calibration is by far faster and more problem oriented than measurements in frequency domain with subsequent Fourier transform processing. Moreover, only few operations are required for each calibration target (2 co- and 2 cross-polarization measurements), compared to the usually hundreds of measurement steps in the frequency domain in order to perform a wideband characterization of the target. Furthermore, the time gating for the elimination of the background reflections and the direct antenna coupling can be directly performed.

In this chapter the UWB Radar link is investigated. This is done by firstly deriving a fully polarimetric description of the UWB Radar link in the time domain. Secondly, in order to characterize different targets in the time domain, the polarimetric components of the target impulse response are calculated in the time domain, through a calibration process. Measurements are presented to validate the proposed approach.

7.1 Derivation of the UWB Radar link in the time domain

In this section a model for the UWB Radar link in the time domain is presented. The analyzed system is composed of a transmitter, a scattering object and a receiver, as illustrated in Fig. 7.1. Let consider dual polarized antennas at the transmitter and at the receiver side.

Let $u_{\text{Tx}}(t)$ be the signal that excites the transmit antenna, which is characterized by the polarimetric antenna impulse response matrix

$$[\mathbf{h}_{\text{Tx}}(t, \theta_{\text{Tx}}, \psi_{\text{Tx}})] = \begin{bmatrix} h_{\text{Tx}}^{\text{hh}}(t, \theta_{\text{Tx}}, \psi_{\text{Tx}}) & h_{\text{Tx}}^{\text{hv}}(t, \theta_{\text{Tx}}, \psi_{\text{Tx}}) \\ h_{\text{Tx}}^{\text{vh}}(t, \theta_{\text{Tx}}, \psi_{\text{Tx}}) & h_{\text{Tx}}^{\text{vv}}(t, \theta_{\text{Tx}}, \psi_{\text{Tx}}) \end{bmatrix} \tag{7.1}$$

where the term h_{Tx}^{ij} indicates the antenna impulse response when the input signal feeds the j polarized channel (where j is either the vertical ^{v} or the horizontal ^{h} polarization) and this signal in the j polarized channel produces a transmitted wave in the i polarization (where i is either the vertically ^{v} or the horizontally ^{h} transmitted wave). This means that a signal feeding the j polarized channel of the antenna produces both the desired j polarized transmitted wave and a i polarized transmitted wave due to antenna coupling.

The input signal $u_{\text{Tx}}(t)$ is in the following considered as a vector

$$[\mathbf{u}_{\text{Tx}}(t)] = \begin{bmatrix} u_{\text{Tx}}^{\text{h}}(t) \\ u_{\text{Tx}}^{\text{v}}(t) \end{bmatrix} \tag{7.2}$$

where each component feeds one polarization (^{h} or ^{v}) of the dual polarized transmit antenna, i.e. with ^{h} the horizontal polarization of the transmit signal is indicated and accordingly for ^{v}. Let a^{h} and a^{v} be the horizontally and vertically transmitted wave components at the antenna output, respectively, i.e.

$$\begin{aligned} \begin{bmatrix} a^{\text{h}}(t, \theta_{\text{Tx}}, \psi_{\text{Tx}}) \\ a^{\text{v}}(t, \theta_{\text{Tx}}, \psi_{\text{Tx}}) \end{bmatrix} &= \frac{1}{2\pi c_0} [\mathbf{h}_{\text{Tx}}(t, \theta_{\text{Tx}}, \psi_{\text{Tx}})] * \frac{1}{\sqrt{Z_{\text{Tx}}}} \frac{\partial}{\partial t} [\mathbf{u}_{\text{Tx}}(t)] \\ &= \frac{1}{2\pi c_0} \frac{1}{\sqrt{Z_{\text{Tx}}}} \begin{bmatrix} h_{\text{Tx}}^{\text{hh}}(t, \theta_{\text{Tx}}, \psi_{\text{Tx}}) * \frac{\partial}{\partial t} u_{\text{Tx}}^{\text{h}}(t) + h_{\text{Tx}}^{\text{hv}}(t, \theta_{\text{Tx}}, \psi_{\text{Tx}}) * \frac{\partial}{\partial t} u_{\text{Tx}}^{\text{v}}(t) \\ h_{\text{Tx}}^{\text{vv}}(t, \theta_{\text{Tx}}, \psi_{\text{Tx}}) * \frac{\partial}{\partial t} u_{\text{Tx}}^{\text{v}}(t) + h_{\text{Tx}}^{\text{vh}}(t, \theta_{\text{Tx}}, \psi_{\text{Tx}}) * \frac{\partial}{\partial t} u_{\text{Tx}}^{\text{h}}(t) \end{bmatrix} . \end{aligned} \tag{7.3}$$

The derivative is caused by the transmit antenna, as already clarified in chapter 4 [46]. The mixed terms ^{hv} and ^{vh} result from the polarization coupling, as previously explained. The symbol $*$ denotes the convolution between the matrix $[\mathbf{h}_{\text{Tx}}]$ and the vector $[\mathbf{u}_{\text{Tx}}]$. This convolution has to be interpreted as a matrix multiplication but with a convolution operation instead of a multiplication operation.

According to chapter 5, the radiated electric field at a distance r is hence

$$[\mathbf{e}_i] = \frac{1}{r} [\mathbf{h}_{\text{Chf}}] * [\mathbf{a}] \tag{7.4}$$

where the index i indicates that the electric field $[\mathbf{e}_i]$ is impinging on the object and the matrix

$$[\mathbf{h}_{\text{Chf}}] = \begin{bmatrix} h_{\text{Chf}}^{\text{hh}} & h_{\text{Chf}}^{\text{hv}} \\ h_{\text{Chf}}^{\text{vh}} & h_{\text{Chf}}^{\text{vv}} \end{bmatrix} \tag{7.5}$$

represents the polarimetric impulse response matrix of the forward channel, i.e. the channel to the target object.

According to Fig. 7.1, the measured environment is composed of the transmit path, the scattering object, the receive path and the unwanted contributions due to the antenna coupling and room background reflections, which are both described by the coupling impulse response $[\mathbf{h}_{\text{coupl}}(t)]$. All these contributions are taken into account by the measured impulse response matrix $[\mathbf{h}_{\text{m}}]$, as symbolically shown in Fig. 7.1. Let b^{h} and b^{v} be the horizontal and vertical wave components at the receive antenna (see Fig. 7.1). In the time domain, the scenario can be written as

$$
\begin{bmatrix} b^{\text{h}} \\ b^{\text{v}} \end{bmatrix} = [\mathbf{h}_{\text{m}}] * \begin{bmatrix} a^{\text{h}} \\ a^{\text{v}} \end{bmatrix}
$$

$$
= \frac{1}{2\pi c_0} \begin{bmatrix} h_{\text{m}}^{\text{hh}} & h_{\text{m}}^{\text{hv}} \\ h_{\text{m}}^{\text{vh}} & h_{\text{m}}^{\text{vv}} \end{bmatrix} * \begin{bmatrix} h_{\text{Tx}}^{\text{hh}} & h_{\text{Tx}}^{\text{hv}} \\ h_{\text{Tx}}^{\text{vh}} & h_{\text{Tx}}^{\text{vv}} \end{bmatrix} * \frac{1}{\sqrt{Z_{\text{Tx}}}} \frac{\partial}{\partial t} [\mathbf{u}_{\text{Tx}}(t)] . \tag{7.6}
$$

As previously said, the matrix $[\mathbf{h}_{\text{m}}]$ contains the transmit path, the scattering object, the receive path, the antenna coupling and the background reflections terms. The antenna coupling is present by coupling between the transmit and receive antennas. The actually received signal vector

$$
[\mathbf{u}_{\text{Rx}}(t)] = \begin{bmatrix} u_{\text{Rx}}^{\text{h}}(t) \\ u_{\text{Rx}}^{\text{v}}(t) \end{bmatrix} \tag{7.7}
$$

with ^{h} the recovered signal from the horizontal and with ^{v} from the vertical polarization of the receive antenna, respectively, is then obtained including in the previous equation also the receive antenna impulse response matrix $[\mathbf{h}_{\text{Rx}}]$, namely [39]

$$
\frac{[\mathbf{u}_{\text{Rx}}(t)]}{\sqrt{Z_{\text{Rx}}}} = \frac{1}{2\pi c_0} [\mathbf{h}_{\text{Rx}}(t, \theta_{\text{Rx}}, \psi_{\text{Rx}})]^{\text{T}} * [\mathbf{h}_{\text{m}}] * [\mathbf{h}_{\text{Tx}}(t, \theta_{\text{Tx}}, \psi_{\text{Tx}})]
$$

$$
* \frac{1}{\sqrt{Z_{\text{Tx}}}} \frac{\partial}{\partial t} [\mathbf{u}_{\text{Tx}}(t)] \tag{7.8}
$$

where the symbol ^{T} indicates the transposed matrix. From the previous equation, it can be evinced that the dimensions of $[\mathbf{h}_{\text{m}}]$ are $1/\text{m}$, since the dimensions of $[\mathbf{h}_{\text{Tx}}]$, $[\mathbf{h}_{\text{Rx}}]$, c_0 are m/s and each convolution has the dimension s. It has to be pointed out that the matrix $[\mathbf{h}_{\text{m}}]$ includes influences due to mismatches in the hardware equipment, coupling between the forward and return channels, and residual reflections in the chamber. Consequently, in order to correctly detect the target impulse response, a calibration of the system is required, so that the unwanted distortions and coupling terms between the antennas and also the background reflections can be cancelled.

7.1.1 The Error Cube in the Time Domain

Let consider, at the beginning, the measurement environment of Fig. 7.1. All propagations effects that are occurring between the transmit antenna and the receive antenna

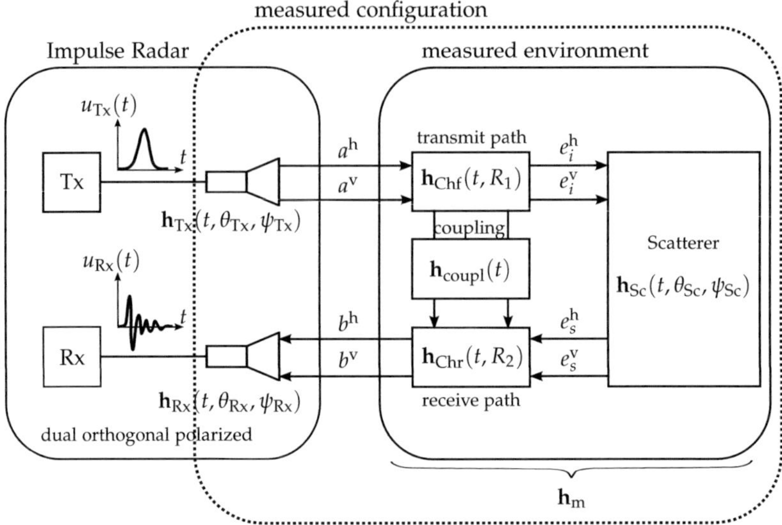

Figure 7.1: The measured configuration

are contained in the matrix $[\mathbf{h}_m]$. The influence of the antennas is included according to eq. (7.8). The time domain relationship between $\mathbf{a} = [a^h, a^v]^T$ and $\mathbf{b} = [b^h, b^v]^T$ can be described by the signal flow error cube for polarimetric antennas which was introduced in [73] in the frequency domain. Starting from that definition, it is now extended to the time domain as illustrated in Fig. 7.2. All signal paths are on the edges or on the diagonals of the cube (see Fig. 7.2(b)). The input and output time domain waves a^h, a^v, b^h and b^v are located at the four front corners of the cube, in the so called Radar plane (see Fig. 7.2(a)). It contains the isolation errors due to the direct antenna coupling and due to the background chamber reflections ("empty room" reflections). These terms can be expressed by a matrix $[\mathbf{h}_{coupl}]$ of the form

$$[\mathbf{h}_{coupl}(t, \theta, \psi)] = \begin{bmatrix} h_{coupl}^{hh}(t, \theta, \psi) & h_{coupl}^{hv}(t, \theta, \psi) \\ h_{coupl}^{vh}(t, \theta, \psi) & h_{coupl}^{vv}(t, \theta, \psi) \end{bmatrix} \tag{7.9}$$

where h_{coupl}^{ij} represents the co-polarized coupling for $i = j$ and the cross-polarized coupling for $i \neq j$.

The top plane of the error cube is called horizontal plane, because it contains the horizontal signal paths. Similarly, the bottom plane, called vertical plane, contains the vertical signal paths. The influence of the cross-polarization terms ^{hv}, ^{vh} is represented by diagonal connections, while edge connections are pure co-polarized terms ^{hh}, ^{vv}.

The forward and return paths, described by the matrices $[\mathbf{h}_{Chf}]$ and $[\mathbf{h}_{Chr}]$, respectively, are located in the sidewalls. The related impulse responses are a function of the distance between the scatterer and the receive antennas. The desired co-polarized terms are located on the edges, while the unwanted cross-polarized terms are located on the diagonals.

Finally, the back wall represents the target plane, where the scatterer's co-polarized impulse response components h_{Sc}^{hh}, h_{Sc}^{vv} are positioned on the edges and the cross-polarized impulse response components h_{Sc}^{hv}, h_{Sc}^{hv} lie on the diagonals. According to this signal path description, the measured system impulse response matrix in the time domain can be written as [73]

$$[\mathbf{h_m}] = [\mathbf{h_{coupl}}] + \frac{1}{r}[\mathbf{h_{Chr}}] * [\mathbf{h_{Sc}}] * \frac{1}{r}[\mathbf{h_{Chf}}] \tag{7.10}$$

where the dependence on the distance r between the target and the antennas (quasi-monostatic case) is explicitly written. The first term $1/r$ is the distance dependence for the return channel and the second term indicates the distance dependence for the forward channel (see eq. (7.4)) [39]. The dimensions of $[\mathbf{h_m}]$ are $1/m$ and since the channels produce time shifts and have the dimension $1/s$, the dimensions of $[\mathbf{h_{Sc}}]$ are m/s (the dimension of the convolution is s). $[\mathbf{h_{Sc}}]$ is defined as the inverse Fourier transform of the object scattering matrix, as it is in detail described in section 7.2. Finally, taking into account also the impulse response matrices of the transmit and of the receive antennas, according to eq. (7.8) the overall time domain input-output relationship can be written as

$$\frac{[\mathbf{u_{Rx}}(t)]}{\sqrt{Z_{Rx}}} = \frac{1}{2\pi}[\mathbf{h_{Rx}}(t,\theta_{Rx},\psi_{Rx})]^T * \Big\{ [\mathbf{h_{coupl}}(t)] +$$

$$\frac{1}{r}[\mathbf{h_{Chr}}(t)] * [\mathbf{h_{Sc}}(t,\theta_{Sc},\psi_{Sc})] * \frac{1}{r}[\mathbf{h_{Chf}}(t)] \Big\}$$

$$* [\mathbf{h_{Tx}}(t,\theta_{Tx},\psi_{Tx})] * \frac{1}{\sqrt{Z_{Tx}}}\frac{\partial}{\partial t}[\mathbf{u_{Tx}}(t)] . \tag{7.11}$$

Observing the previous equation, it can be inferred that, in order to directly measure the scatterer's impulse response, the unknowns in eq. (7.11) have to be determined.

7.2 UWB Radar Targets

In many Radar applications, the aim is to detect a particular target and to classify it according to information contained in the signal back-scattered by the target itself. This classification can be performed in the frequency domain or in the time domain. The quantities to be considered are different in these two cases, and also the resulting information has different physical meaning. In the following, a characterization of targets for UWB Radar applications is presented, both in the frequency domain (which is the usual characterization) and in the time domain.

7.2.1 Frequency Domain Description

Let $[\mathbf{E}_i]$ be the field incident on the target and $[\mathbf{E}_s]$ the scattered field from the target at a distance r from the target. It is usual to define a polarization scattering matrix $[\mathbf{S}]$ for

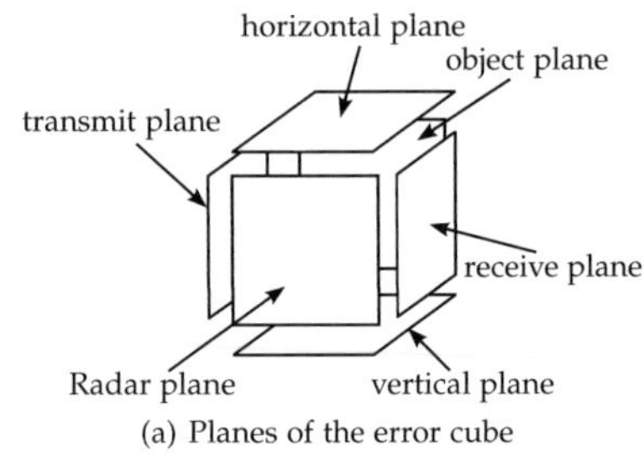

(a) Planes of the error cube

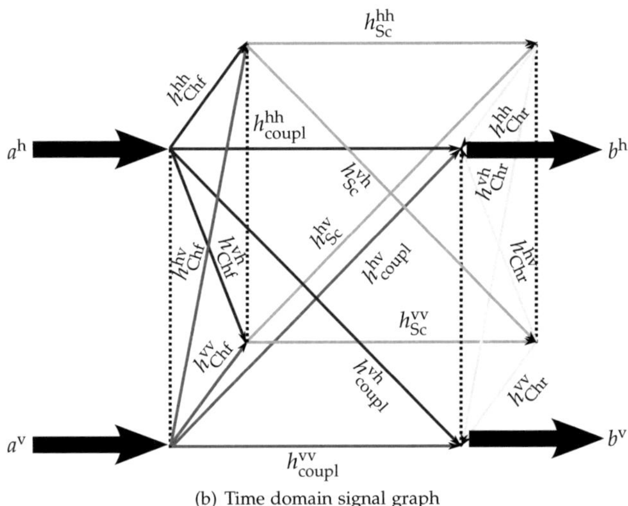

(b) Time domain signal graph

Figure 7.2: Error cube in the time domain.

the object in the following way [75]

$$[\mathbf{E}_s] = [\mathbf{S}][\mathbf{E}_i] \tag{7.12}$$

where $\mathbf{S}$ is dimensionless. It has to be pointed out that the matrix $[\mathbf{S}]$ depends on the distance r at which the electric field is recovered. Hence, in order to eliminate the dependence on the observation range, the polarization scattering matrix is defined as the Sinclair matrix [76], namely

$$[\mathbf{E}_s] = \frac{1}{\sqrt{4\pi r^2}}[\mathbf{S}_{Sc}][\mathbf{E}_i]$$

$$\begin{bmatrix} E_s^h \\ E_s^v \end{bmatrix} = \frac{1}{\sqrt{4\pi r^2}} \begin{bmatrix} S_{Sc}^{hh} & S_{Sc}^{hv} \\ S_{Sc}^{vh} & S_{Sc}^{vv} \end{bmatrix} \begin{bmatrix} E_i^h \\ E_i^v \end{bmatrix} \tag{7.13}$$

i.e. it coincides with the previous one in eq (7.12), divided by the distance r. The matrix defined in (7.13) contains the same information as the polarization scattering matrix defined in (7.12) but it is not dependent on the observation range r, i.e. it is object oriented [75]. The dimensions of $[\mathbf{S}_{Sc}]$ defined in (7.13) are hence m. Hereafter, the definition in (7.13) is adopted.

In the frequency domain the scattering behavior of the target is described by its Radar Cross Section (RCS) $[\sigma_{Sc}(f)]$. The fully polarimetric scattering behavior is expressed by a 2×2 matrix as

$$[\sigma_{Sc}(f)] = \begin{bmatrix} \sigma_{Sc}^{hh}(f) & \sigma_{Sc}^{hv}(f) \\ \sigma_{Sc}^{vh}(f) & \sigma_{Sc}^{vv}(f) \end{bmatrix} = 4\pi[\mathbf{S}_{Sc}^2(f)] = \begin{bmatrix} \left(S_{Sc}^{hh}(f)\right)^2 & \left(S_{Sc}^{hv}(f)\right)^2 \\ \left(S_{Sc}^{vh}(f)\right)^2 & \left(S_{Sc}^{vv}(f)\right)^2 \end{bmatrix} \tag{7.14}$$

where the matrix $[\mathbf{S}_{Sc}(f)]$ is the polarization scattering matrix of the target defined in (7.13). Since the scattering matrix coincides with the frequency dependent transfer function matrix $[\mathbf{H}_{Sc}(f)]$, it results that $[\sigma_{Sc}(f)] = 4\pi[\mathbf{H}_{Sc}^2(f)]$. The absolute value of each component $|\sigma_{Sc}^{ij}(f)|$ of the RCS matrix, where ^{ij} expresses one particular polarimetric polarization (i.e. ^{hh}, ^{hv}, ^{vh}, ^{vv}), is directly proportional to the squared absolute value of the transfer function $|H_{Sc}^{ij}(f)|^2$ of the target and hence it describes the power distribution in the frequency domain.

7.2.2 Time Domain Description

Let $[\mathbf{h}_{Sc}(t)]$ denote the impulse response matrix of the scatterer, defined as

$$[\mathbf{h}_{Sc}(t)] = \begin{bmatrix} h_{Sc}^{hh}(t) & h_{Sc}^{hv}(t) \\ h_{Sc}^{vh}(t) & h_{Sc}^{vv}(t) \end{bmatrix} = \text{IFT}\,\{[\mathbf{H}_{Sc}(f)]\}$$

$$= \begin{bmatrix} \text{IFT}\,\{H_{Sc}^{hh}(f)\} & \text{IFT}\,\{H_{Sc}^{hv}(f)\} \\ \text{IFT}\,\{H_{Sc}^{vh}(f)\} & \text{IFT}\,\{H_{Sc}^{vv}(f)\} \end{bmatrix} \tag{7.15}$$

where IFT denotes the Inverse Fourier Transform.

From the frequency domain equivalence between the scattering matrix and the transfer function matrix, namely $[\mathbf{S}_{Sc}(f)] = [\mathbf{H}_{Sc}(f)]$, it turns out that the time domain equivalent $[\sigma_{Sc}(t)]$ of the frequency domain RCS matrix $[\sigma_{Sc}(f)]$ is given by

$$[\sigma_{Sc}(t)] = 4\pi \begin{bmatrix} \left[\text{IFT}\left\{H_{Sc}^{hh}(f)\right\}\right]^2 & \left[\text{IFT}\left\{H_{Sc}^{hv}(f)\right\}\right]^2 \\ \left[\text{IFT}\left\{H_{Sc}^{vh}(f)\right\}\right]^2 & \left[\text{IFT}\left\{H_{Sc}^{vv}(f)\right\}\right]^2 \end{bmatrix}$$

$$= 4\pi[\mathbf{h}_{Sc}^2(t)] = \begin{bmatrix} \left[h_{Sc}^{hh}(t)\right]^2 & \left[h_{Sc}^{hv}(t)\right]^2 \\ \left[h_{Sc}^{vh}(t)\right]^2 & \left[h_{Sc}^{vv}(t)\right]^2 \end{bmatrix} \tag{7.16}$$

The absolute value of each term $\sigma_{Sc}^{ij}(t)$ is directly proportional to the scatterer's power delay profile $|h_{Sc}^{ij}(t)|^2$ and hence it describes the power distribution in the time domain.

It has to be noticed that, in order to characterize the target completely in the time domain, it is necessary to know its impulse response matrix. Consequently, a calibration procedure in the time domain has to be performed. Hence, from this observation it can be evinced that, directly calibrating the recovered back-scattered signal from a UWB Radar target allows for directly classifying the target in the time domain.

7.2.3 Time Domain Correlation Properties of UWB Radar Scattered Signals

As previously introduced, once the impulse responses of a target have been calibrated for each polarization, it is possible to classify the target itself based on its signature. Classical methods for classifying targets are based on frequency domain signatures [41], [77]. In the following, a time domain method is introduced. This method is based on the correlation of the signal backscattered from the target with a reference template.

Let $h_{Sc,id}^{ij}$ be the ideally expected target's impulse response for the ^{ij} polarization and let h_{Sc}^{ij} be the target's impulse response for the same polarization deconvolved from a calibrated measurement. According to the theory developed in chapter 2 (ref. to section 2.3), through the calculation of the fidelity between $h_{Sc,id}^{ij}$ and h_{Sc}^{ij} it is possible to determine the amount of distortion between the two impulse responses [78]. This procedure has to be performed for each polarization component.

Hence, by constructing a database containing the information about the ideally expected impulse response for each polarization for different targets, by calculating the fidelity with the measured calibrated impulse response, it is possible to determine the unknown target.

7.3 Fully Polarimetric UWB Time Domain Radar Calibration

As previously stated, in order to correctly detect and classify the target, a calibration of the system shown in Fig. 7.1 is necessary, so that the contributions of the unwanted terms in eq. (7.11) like coupling, background reflections, etc. can be cancelled. In this section, a complete procedure for calibrating the UWB Radar system in the time domain is presented. The importance of performing this calibration directly in the time domain lies on the fact that with this procedure it is directly possible to assess the time domain behavior of the targets and hence to directly classify them according to their time domain signature. Obtaining the calibrated scatterer's impulse response with a procedure directly performed in the time domain permits also to quantify the distribution of the power in time.

Additionally, in the time domain only few measurements of impulse responses are required, compared to hundreds of frequency steps in a wideband frequency domain measurement. Moreover, an additional time gating can be directly performed, which permits to discern between the wanted and unwanted contributions (coupling, reflections, …) based on their time on arrival.

The first step to recover the scatterer's impulse response matrix $[h_{Sc}]$, starting from the measured received voltage vector $[u_{Rx}]$, consists in deconvolving the transmit and the receive antennas' impulse response matrices from eq. (7.11). In the following, the calibration of the measured environment (Fig. 7.1) is investigated, assuming that the matrices $[h_{Tx}]$ and $[h_{Rx}]$ are already known. These matrices can be obtained e.g. with the measurement procedures described in chapter 4.

7.3.1 Measurement Setup

For a time domain UWB Radar calibration, the data are directly acquired in the time domain. The measurement system is a quasi-monostatic Radar system. It is composed of a transmit antenna and a receive antenna, positioned very near to each other. At the transmitter side a pulse generator creates a baseband UWB pulse, which is fed to the transmit antenna. At the receiver side, the signal recovered by the receive antenna is sampled, which is usually performed with a high-speed realtime oscilloscope. A trigger signal synchronizes both the pulse generator and the oscilloscope. A scatterer is positioned at a distance r from the two antennas.

7.3.2 Empty Room Calibration

In the typical frequency domain calibration procedure, for the determination of the coupling and the background reflections, i.e. $[h_{coupl}]$ in the measured matrix $[h_m]$, a measurement of the empty room is performed. In the time domain this calibration step is not necessary anymore if time gating is applied, as it is now explained.

In the "empty room" case there is no target and hence $[h_{Sc}] = [0]$. Consequently, the measured impulse response matrix $[h_m]$ results in

$$[h_m] = [h_{coupl}] + \frac{1}{r}[h_{Chr}] * \underbrace{[h_{Sc}]}_{=[0]} * \frac{1}{r}[h_{Chf}] = [h_{coupl}] \qquad (7.17)$$

Hence, the contributions of the "empty room" are given by the matrix $[h_{coupl}]$ only, which includes the coupling between the antennas and the background reflections.

In Fig. 7.3 the signal $u_{\mathrm{Rx}}^{\mathrm{h}}$ recovered by the oscilloscope for the ^{hh} polarization is plotted: (top) in the case of the empty room and (bottom) with a scatterer (flat plate). The delay of about 10 ns for the first high peak is due to the delay introduced by the connecting cables. In the complete processing algorithm (see section 7.4) this delay is compensated. From the plot of the measured received signal for the "empty room" in Fig. 7.3(a) it can be recognized that the recovered signal has a very high peak (approx. 0.015 V) after approx. 12 ns. The high peak is caused by the coupling between the transmit and the receive antenna. In addition very small peaks at around 40 ns are visible, which occur due to the reflection from the back wall of the anechoic chamber. On the other hand, the presence of a scatterer positioned at a distance r from the antennas produces a peak in the received signal at a time $t = \tilde{r}/c_0 + t_{\mathrm{A}}$, where $\tilde{r}$ takes into account also of the length of the connecting cables and t_{A} of the delay introduced by the antennas. This is shown in Fig. 7.3(b) for a flat plate, ^{hh} polarization. It can be recognized that the high peak caused by the plate can be separated in the time domain from the direct coupling contribution and from the backwall reflections, i.e. the time of arrival of the unwanted contributions of the antenna coupling and of the background reflections are different with respect to the time of arrival of the contribution of the scatterer. In the following time gating will be applied and hence the empty room measurement is not necessary anymore in the time domain UWB Radar calibration procedure.

7.3.3 Channel Calibration

Rewriting eq. (7.10) by expanding the convolution operations, the measured system response polarimetric components result

$$
\begin{bmatrix} \tilde{h}_{\mathrm{m}}^{\mathrm{hh}} \\ \tilde{h}_{\mathrm{m}}^{\mathrm{hv}} \\ \tilde{h}_{\mathrm{m}}^{\mathrm{vh}} \\ \tilde{h}_{\mathrm{m}}^{\mathrm{vv}} \end{bmatrix} = \underbrace{\begin{bmatrix} h_{\mathrm{Chr}}^{\mathrm{hh}} * h_{\mathrm{Chf}}^{\mathrm{hh}} & h_{\mathrm{Chr}}^{\mathrm{hh}} * h_{\mathrm{Chf}}^{\mathrm{vh}} & h_{\mathrm{Chr}}^{\mathrm{hv}} * h_{\mathrm{Chf}}^{\mathrm{hh}} & h_{\mathrm{Chr}}^{\mathrm{hv}} * h_{\mathrm{Chf}}^{\mathrm{vh}} \\ h_{\mathrm{Chr}}^{\mathrm{hh}} * h_{\mathrm{Chf}}^{\mathrm{hv}} & h_{\mathrm{Chr}}^{\mathrm{hh}} * h_{\mathrm{Chf}}^{\mathrm{vv}} & h_{\mathrm{Chr}}^{\mathrm{hv}} * h_{\mathrm{Chf}}^{\mathrm{hv}} & h_{\mathrm{Chr}}^{\mathrm{hv}} * h_{\mathrm{Chf}}^{\mathrm{vv}} \\ h_{\mathrm{Chr}}^{\mathrm{vh}} * h_{\mathrm{Chf}}^{\mathrm{hh}} & h_{\mathrm{Chr}}^{\mathrm{vh}} * h_{\mathrm{Chf}}^{\mathrm{vh}} & h_{\mathrm{Chr}}^{\mathrm{vv}} * h_{\mathrm{Chf}}^{\mathrm{hh}} & h_{\mathrm{Chr}}^{\mathrm{vv}} * h_{\mathrm{Chf}}^{\mathrm{vh}} \\ h_{\mathrm{Chr}}^{\mathrm{vh}} * h_{\mathrm{Chf}}^{\mathrm{hv}} & h_{\mathrm{Chr}}^{\mathrm{vh}} * h_{\mathrm{Chf}}^{\mathrm{vv}} & h_{\mathrm{Chr}}^{\mathrm{vv}} * h_{\mathrm{Chf}}^{\mathrm{hv}} & h_{\mathrm{Chr}}^{\mathrm{vv}} * h_{\mathrm{Chf}}^{\mathrm{vv}} \end{bmatrix}}_{[\mathbf{C}]} *
$$

$$
\begin{bmatrix} h_{\mathrm{Sc}}^{\mathrm{hh}} \\ h_{\mathrm{Sc}}^{\mathrm{hv}} \\ h_{\mathrm{Sc}}^{\mathrm{vh}} \\ h_{\mathrm{Sc}}^{\mathrm{vv}} \end{bmatrix} \tag{7.18}
$$

where $\tilde{h}_{\mathrm{m}}^{ij}$ indicates the time-gated version of h_{m}^{ij}, i.e. after the elimination by time gating of the coupling contribution and of the background reflections, as explained in the previous section, and the commutativity property of the convolution has been used.

In order to simplify the notation, let the element in position ij in the previously defined matrix $[\mathbf{C}]$ be indicated as c_{ij}. For carrying out the calibration, the terms c_{ij} of the matrix $[\mathbf{C}]$ have to be determined. Measurements with reference targets are needed to determine the 16 unknowns in $[\mathbf{C}]$. In the following a particular procedure

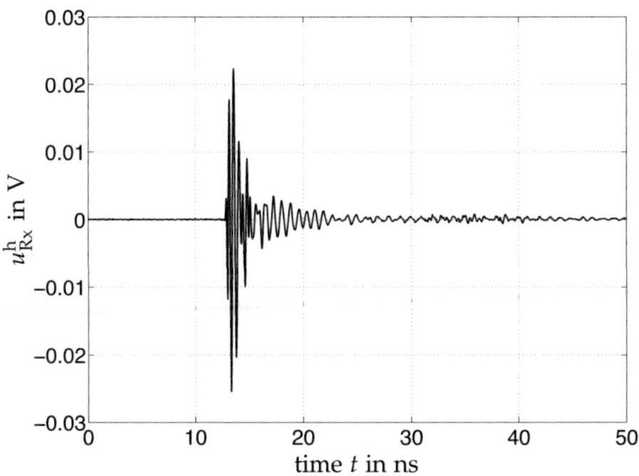

(a) Measured $u_{\mathrm{Rx}}^{\mathrm{h}}$ for the empty room (for ^{hh} polarization).

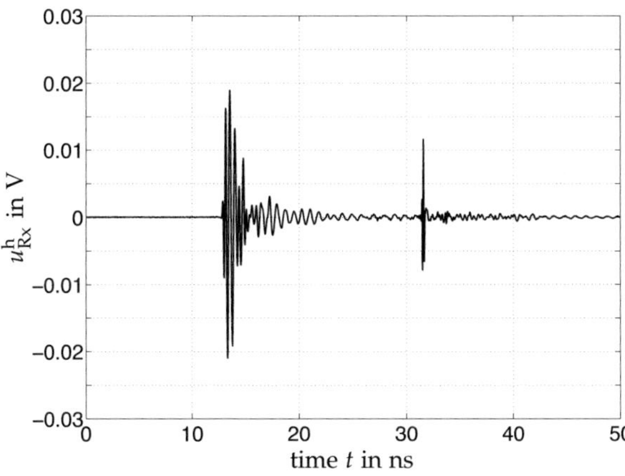

(b) Measured $u_{\mathrm{Rx}}^{\mathrm{h}}$ for a flat plate (for ^{hh} polarization).

Figure 7.3: Measured received signal $u_{\mathrm{Rx}}^{\mathrm{h}}$ (for ^{hh} polarization) without scatterer (a) and with scatterer (b).

is presented, based on [73], [75]. It can be observed that not all terms of the matrix $[\mathbf{C}]$ are linearly independent. It can be shown that the entire matrix can be described by 8 elements.

As an example, let consider $c_{41} = h^{vh}_{Chr} * h^{hv}_{Chf}$. It can be written as

$$h^{vh}_{Chr} * h^{hv}_{Chf} = h^{vh}_{Chr} * \underbrace{h^{hh}_{Chf} *^{-1} h^{hh}_{Chf}}_{\delta} * \underbrace{h^{hh}_{Chr} *^{-1} h^{hh}_{Chr}}_{\delta} * h^{hv}_{Chf}$$

$$= h^{vh}_{Chr} * h^{hh}_{Chf} * h^{hh}_{Chr} * h^{hv}_{Chf} *^{-1} h^{hh}_{Chr} *^{-1} h^{hh}_{Chf}$$

$$= \underbrace{h^{vh}_{Chr} * h^{hh}_{Chf}}_{c_{31}} * \underbrace{h^{hh}_{Chr} * h^{hv}_{Chf}}_{c_{21}} *^{-1} \Bigg(\underbrace{h^{hh}_{Chr} * h^{hh}_{Chf}}_{c_{11}} \Bigg) \tag{7.19}$$

where the symbol $a(t) *^{-1} b(t)$ indicates the deconvolution between $a(t)$ and $b(t)$ and the deconvolution property $a(t) *^{-1} a(t) = \delta(t)$ has been used (in eq. (7.19) the time-dependence of the various functions and of $\delta(t)$ has been omitted for notation simplification).

Doing the same procedure for the other matrix elements c_{ij}, eq. (7.18) can be simplified as

$$[\mathbf{C}] =$$

$$\begin{bmatrix} c_{11} & c_{34} * c_{11} *^{-1} c_{33} & c_{33} * c_{24} *^{-1} c_{44} & c_{34} * c_{24} *^{-1} c_{44} \\ c_{21} & c_{22} & c_{21} * c_{24} *^{-1} c_{22} & c_{24} \\ c_{31} & c_{31} * c_{34} *^{-1} c_{33} & c_{33} & c_{34} \\ c_{31} * c_{21} *^{-1} c_{11} & c_{31} * c_{44} *^{-1} c_{33} & c_{33} * c_{21} *^{-1} c_{11} & c_{44} \end{bmatrix} \tag{7.20}$$

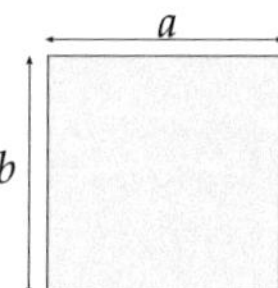

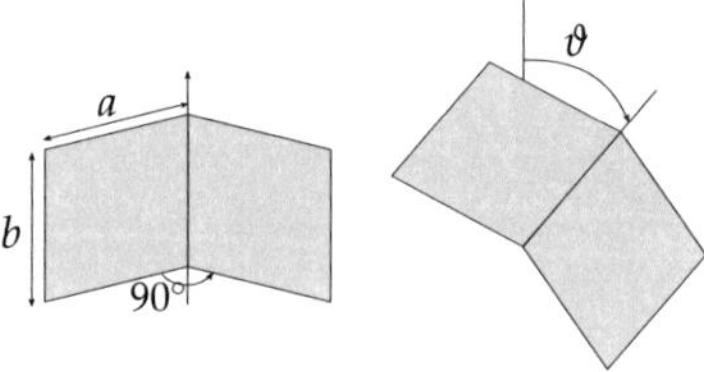

Figure 7.4: Polarimetric calibration targets: flat plate (top) and vertical dihedral with $0°$ offset (bottom left) and with ϑ offset (bottom right) from the vertical.

7.3.3.1 Choice of the Polarimetric Calibration Targets

It can be shown that under the hypotheses of linear independency between the scattering matrices of the selected reference targets and reciprocity of each scattering matrix, the 8 linearly independent unknown parameters c_{ij} can be derived using only three reference targets. Let the polarization scattering matrix of the target n (with $n = 1, 2, 3$) be $[S_n]$, as defined in eq. (7.13). Using the reciprocity condition, which means that $S_n^{vh} = S_n^{hv}$, it can be written

$$[S_n] = \begin{bmatrix} S_n^{hh} & S_n^{hv} \\ S_n^{vh} & S_n^{vv} \end{bmatrix} \underset{\substack{\uparrow \\ \text{reciprocity}}}{=} \begin{bmatrix} S_n^{hh} & S_n^{hv} \\ S_n^{hv} & S_n^{vv} \end{bmatrix}. \tag{7.21}$$

The linear independency of the scattering matrices of the three different reference targets to be chosen assures that the matrix $[\tilde{S}]$ constructed in the following way

$$[\tilde{S}] = \begin{bmatrix} S_1^{vv} & S_1^{hh} & S_1^{hv} \\ S_2^{vv} & S_2^{hh} & S_2^{hv} \\ S_3^{vv} & S_3^{hh} & S_3^{hv} \end{bmatrix} \tag{7.22}$$

has full rank, i.e. its determinant is non-zero. The selected set of targets, which satisfies the given conditions, is the following

Flat Plate: The scattering matrix of a flat plate has only co-polarized components and they are equal in amplitude and in-phase. The scattering matrix (as defined in eq. (7.13)) and the related impulse response matrix of a plate (positioned orthogonal to the transmit antenna, i.e. normal incidence), are given by [76], [79]

$$[S_{\text{plate}}(f)] = |S_{\text{plate}}(f)| \begin{bmatrix} 1 & 0 \\ 0 & 1 \end{bmatrix} \circ\!\!-\!\!\bullet \; [h_{\text{plate}}(t)] = |h_{\text{plate}}(t)| \begin{bmatrix} 1 & 0 \\ 0 & 1 \end{bmatrix} \tag{7.23}$$

where $h_{\text{plate}}(t)$ is the inverse Fourier transform of $S_{\text{plate}}(f)$. According to [76], $S_{\text{plate}}(f)$ can be analytically calculated as (at normal incidence)

$$S_{\text{plate}}(f) = 2\sqrt{\pi} A_{\text{plate}} \frac{f}{c_0} \tag{7.24}$$

where $A_{\text{plate}} = a \cdot b$ (ref. to Fig. 7.4) is the area of the plate in squared meters. The dimensions of S_{plate} are m and hence the dimension of h_{plate} are m/s, as already previously derived in eq. (7.8) and (7.10).

Dihedral corner reflector, offset $0°$**:** The dihedral reflector, vertically positioned, i.e. with an offset of $0°$ with respect to the vertical, has only co-polarized components, which are equal in amplitude but out-of-phase. The scattering matrix and the related impulse response matrix are given by

$$[S_{\text{Di},0°}(f)] = |S_{\text{Di},0°}(f)| \begin{bmatrix} 1 & 0 \\ 0 & -1 \end{bmatrix} \circ\!\!-\!\!\bullet \; [h_{\text{Di},0°}(t)] = |h_{\text{Di},0°}(t)| \begin{bmatrix} 1 & 0 \\ 0 & -1 \end{bmatrix} \tag{7.25}$$

with $h_{\text{Di},0°}(t)$ being the inverse Fourier transform of $S_{\text{Di},0°}(f)$. According to [76], for a vertically positioned dihedral corner reflector with the incident ray path perpendicular to the dihedral fold line, it is obtained

$$S_{\text{Di},0°}(f) = 4\sqrt{\pi}ab \sin\left(\frac{\pi}{4}\right)\frac{f}{c_0} \tag{7.26}$$

where a and b are the dimensions of the dihedral reflector (see Fig. 7.4).

Dihedral corner reflector, offset $45°$**:** The dihedral reflector, positioned with an offset of $\vartheta = 45°$ with respect to the vertical, has only cross-polarized components, which are equal in amplitude and in-phase. The scattering matrix is related to the one in the case of zero-offset from the vertical and the general solution dependent on the offset angle ϑ is [76]

$$[\mathbf{S}_{\text{Di},\vartheta}(f)] = |S_{\text{Di},0°}(f)|\begin{bmatrix} -\cos(2\vartheta) & \sin(2\vartheta) \\ \sin(2\vartheta) & \cos(2\vartheta) \end{bmatrix} \tag{7.27}$$

For $\vartheta = 45°$, the scattering matrix and the related impulse response matrix are hence given by

$$[\mathbf{S}_{\text{Di},45°}(f)] = |S_{\text{Di},0°}(f)|\begin{bmatrix} 0 & 1 \\ 1 & 0 \end{bmatrix} \circ\!\!-\!\!\bullet [\mathbf{h}_{\text{Di},45°}(t)] = |h_{\text{Di},0°}(t)|\begin{bmatrix} 0 & 1 \\ 1 & 0 \end{bmatrix} \tag{7.28}$$

with $S_{\text{Di},0°}(f) = S_{\text{Di},45°}(f)$ and $h_{\text{Di},45°}(t)$ is the inverse Fourier transform of $S_{\text{Di},45°}(f)$.

7.3.3.2 Channel Calibration Procedure

Once the three reference targets satisfying the two stated conditions (linear independence of the scattering matrices and reciprocity) have been selected, three fully polarimetric measurements, one per target, have to be performed. Taking into account the reciprocity condition, eq. (7.18) can be rewritten after the measurement of target n as

$$\begin{bmatrix} \tilde{h}^{\text{hh}}_{m,n} \\ \tilde{h}^{\text{hv}}_{m,n} \\ \tilde{h}^{\text{vh}}_{m,n} \\ \tilde{h}^{\text{vv}}_{m,n} \end{bmatrix} = \begin{bmatrix} c_{11} * h^{\text{hh}}_{\text{Sc},n} & c_{12} * h^{\text{hv}}_{\text{Sc},n} & c_{13} * h^{\text{vh}}_{\text{Sc},n} & c_{14} * h^{\text{vv}}_{\text{Sc},n} \\ c_{21} * h^{\text{hh}}_{\text{Sc},n} & c_{22} * h^{\text{hv}}_{\text{Sc},n} & c_{23} * h^{\text{vh}}_{\text{Sc},n} & c_{24} * h^{\text{vv}}_{\text{Sc},n} \\ c_{31} * h^{\text{hh}}_{\text{Sc},n} & c_{32} * h^{\text{hv}}_{\text{Sc},n} & c_{33} * h^{\text{vh}}_{\text{Sc},n} & c_{34} * h^{\text{vv}}_{\text{Sc},n} \\ c_{41} * h^{\text{hh}}_{\text{Sc},n} & c_{42} * h^{\text{hv}}_{\text{Sc},n} & c_{43} * h^{\text{vh}}_{\text{Sc},n} & c_{44} * h^{\text{vv}}_{\text{Sc},n} \end{bmatrix}$$

$$= \begin{bmatrix} c_{11} * h^{\text{hh}}_{\text{Sc},n} & (c_{12} + c_{13}) * h^{\text{hv}}_{\text{Sc},n} & c_{14} * h^{\text{vv}}_{\text{Sc},n} \\ c_{21} * h^{\text{hh}}_{\text{Sc},n} & (c_{22} + c_{23}) * h^{\text{hv}}_{\text{Sc},n} & c_{24} * h^{\text{vv}}_{\text{Sc},n} \\ c_{31} * h^{\text{hh}}_{\text{Sc},n} & (c_{32} + c_{33}) * h^{\text{hv}}_{\text{Sc},n} & c_{34} * h^{\text{vv}}_{\text{Sc},n} \\ c_{41} * h^{\text{vv}}_{\text{Sc},n} & (c_{42} + c_{43}) * h^{\text{hv}}_{\text{Sc},n} & c_{44} * h^{\text{vv}}_{\text{Sc},n} \end{bmatrix}$$

$$= \underbrace{\begin{bmatrix} c_{11} & (c_{12} + c_{13}) & c_{14} \\ c_{21} & (c_{22} + c_{23}) & c_{24} \\ c_{31} & (c_{32} + c_{33}) & c_{34} \\ c_{41} & (c_{42} + c_{43}) & c_{44} \end{bmatrix}}_{[\tilde{\mathbf{C}}]} * \begin{bmatrix} h^{\text{hh}}_{\text{Sc},n} \\ h^{\text{hv}}_{\text{Sc},n} \\ h^{\text{vv}}_{\text{Sc},n} \end{bmatrix} \tag{7.29}$$

where the terms $\tilde{h}^{ij}_{m,n}$ are the time-gated versions of $h^{ij}_{m,n}$, as previously explained. Based on reciprocity, it is possible to reduce the 4×4 matrix $[\mathbf{C}]$ of eq. (7.18) to a 4×3 matrix $[\tilde{\mathbf{C}}]$, whose first and last column corresponds to the first and the last column of $[\mathbf{C}]$; the second column is given by the terms $\tilde{c}_{i2} = c_{i2} + c_{i3}$, i.e. it is a linear combination of the second and third columns of $[\mathbf{C}]$.

The equations of (7.29) are resolved for each target. The procedure is now illustrated for the ^{hh} component, being analog for the other components (^{hv}, ^{vh}, ^{vv}). Taking the measured ^{hh} term for each target gives

$$\begin{bmatrix} \tilde{h}^{hh}_{m,1} \\ \tilde{h}^{hh}_{m,2} \\ \tilde{h}^{hh}_{m,3} \end{bmatrix} = \begin{bmatrix} h^{hh}_{Sc,1} & h^{hv}_{Sc,1} & h^{vv}_{Sc,1} \\ h^{hh}_{Sc,2} & h^{hv}_{Sc,2} & h^{vv}_{Sc,2} \\ h^{hh}_{Sc,3} & h^{hv}_{Sc,3} & h^{vv}_{Sc,3} \end{bmatrix} * \begin{bmatrix} c_{11} \\ \tilde{c}_{12} \\ c_{14} \end{bmatrix} = \begin{bmatrix} h_{plate} & 0 & h_{plate} \\ h_{Di,0°} & 0 & -h_{Di,0°} \\ 0 & h_{Di,45°} & 0 \end{bmatrix} * \begin{bmatrix} c_{11} \\ \tilde{c}_{12} \\ c_{14} \end{bmatrix} \qquad (7.30)$$

where the impulse response parameters of the selected targets are used.

Analyzing the previous system of equations, by convolving the term h_{plate} with the second row and deconvolving the term $h_{Di,0°}$ from the same row and then adding the modified second row to the first one, it results

$$\tilde{h}^{hh}_{m,1} + h_{plate} * h_{Di,0°} *^{-1} \tilde{h}^{hh}_{m,2} = h_{plate} * c_{11} + h_{plate} * \underbrace{h_{Di,0°} *^{-1} h_{Di,0°}}_{\delta} * c_{14}$$

$$+ h_{plate} * \underbrace{h_{Di,0°} *^{-1} h_{Di,0°}}_{\delta} * c_{11}$$

$$- h_{plate} * \underbrace{h_{Di,0°} *^{-1} h_{Di,0°}}_{\delta} * c_{14}$$

$$= 2h_{plate} * c_{11} \qquad (7.31)$$

where the symbol $*^{-1}$ indicates the deconvolution operator and the time dependence of the various functions and of $\delta(t)$ has been neglected for notation simplification.

Since h_{plate} can be analytically calculated (ref. to eq. (7.24)), through the deconvolution of h_{plate} from the previous equation, the parameter c_{11} can be obtained, namely

$$c_{11} = \frac{1}{2} h_{plate} *^{-1} \left(\tilde{h}^{hh}_{m,1} + h_{plate} * h_{Di,0°} *^{-1} \tilde{h}^{hh}_{m,2} \right) \qquad (7.32)$$

Analogously, the other terms c_{j1}, $j = 2,3$, can be derived. The term c_{41} is obtained according to the relation in eq. (7.20).

By rewriting eq. (7.30) for the ^{hv} measurements, namely

$$\begin{bmatrix} \tilde{h}^{hv}_{m,1} \\ \tilde{h}^{hv}_{m,2} \\ \tilde{h}^{hv}_{m,3} \end{bmatrix} = \begin{bmatrix} h_{plate} & 0 & h_{plate} \\ h_{Di,0°} & 0 & -h_{Di,0°} \\ 0 & h_{Di,45°} & 0 \end{bmatrix} * \begin{bmatrix} c_{21} \\ \tilde{c}_{22} \\ c_{24} \end{bmatrix} \qquad (7.33)$$

it can be recognized that the term c_{24} can be obtained by deconvolving h_{plate} and convolving $h_{Di,0°}$ to the second row and then subtracting the modified second row from the first one. It results

$$c_{24} = \frac{1}{2} h_{Di,0°} *^{-1} \left(\tilde{h}^{hv}_{m,1} - h_{plate} *^{-1} h_{Di,0°} * \tilde{h}^{hv}_{m,2} \right) \qquad (7.34)$$

Analogously for c_{34} and c_{44}.

The terms $\tilde{c}_{k2}$, with $k = 1, 2, 3, 4$ are obtained from the deconvolution of $h_{\text{Di},45°}$ from $\tilde{h}_{\text{m},2}^{ij}$.

Furthermore, in order to determine c_{43}, let consider the following identity, derived from (7.18) and from the definition of c_{43}

$$
\begin{aligned}
c_{43} &= h_{\text{Chr}}^{\text{vh}} * h_{\text{Chf}}^{\text{vv}} \\
&= \underbrace{h_{\text{Chr}}^{\text{vv}} * h_{\text{Chf}}^{\text{vv}}}_{c_{44}} * \underbrace{h_{\text{Chr}}^{\text{vh}} * h_{\text{Chr}}^{\text{hv}}}_{c_{41}} *^{-1} \underbrace{\left(h_{\text{Chr}}^{\text{vv}} * h_{\text{Chr}}^{\text{hv}} \right)}_{c_{42}}
\end{aligned}
\tag{7.35}
$$

Being $c_{42} = \tilde{c}_{42} - c_{43}$, it results

$$
c_{43} = c_{44} * c_{41} *^{-1} \left(\tilde{c}_{42} - c_{43} \right)
\tag{7.36}
$$

From (7.33), it results that

$$
\tilde{c}_{42} = \tilde{h}_{\text{m},3}^{\text{vv}} *^{-1} h_{\text{Di},45°}.
\tag{7.37}
$$

By reorganizing the terms of equation (7.35) and using (7.37), it is obtained

$$
\begin{aligned}
c_{43} &= c_{44} * c_{41} *^{-1} \tilde{c}_{42} * \left(c_{44} * c_{41} - \delta \right) \\
&= c_{44} * c_{41} *^{-1} \tilde{h}_{\text{m},2}^{\text{vv}} *^{-1} h_{\text{Di},45°} * \left(c_{44} * c_{41} - \delta \right) .
\end{aligned}
\tag{7.38}
$$

Similarly the other parameters c_{2j} and c_{3j}, $j = 1, 2, 3$ can be obtained.

7.3.4 Antenna Calibration

In order to complete the calibration in the time domain, the influence of the transmit and of the receive antennas has to be eliminated. For this purpose the impulse response matrices of the transmit antenna $[\mathbf{h}_{\text{Tx}}]$ and of the receive antenna $[\mathbf{h}_{\text{Rx}}]$ have to be determined. This can be done by the two antennas method, as described in chapter 4 (ref. to section 4.1.1.1). With the knowledge of these two matrices, by using eq. (7.11), the terms in $[\mathbf{h}_{\text{Tx}}]$ and $[\mathbf{h}_{\text{Rx}}]$ are deconvolved from the measured data.

7.4 Verification

In order to validate the proposed time domain calibration for UWB Radar, measurements have been performed.

In this section the system configuration for the verification measurements is presented. Fig. 7.5 shows both a block scheme and a photo of the laboratory facilities during the measurement campaign.

At the transmitter side the system setup is composed of a pulse generator (Pico Second Pulse Lab PSPL 3600), whose output signal u_{Tx} feeds the ^{h} or the ^{v} polarization of the transmit antenna. The signal u_{Tx} has already been shown in Fig. 6.4. The pulse generator is triggered by a rectangular signal generator (Tektronix, 2 GHz Function/Arbitrary Waveform Generator).

At the receiver side the system is composed by the receive antenna and an oscilloscope (Agilent Infinium DCA, 40 GSa/s, dynamic range of 12 bit, 12 GHz bandwidth) that samples the recovered signal u_{Rx} in a single-slot real time acquisition. The oscilloscope is synchronized to the pulse generator by the trigger signal. The system is controlled via a laptop, which permits data acquisition. The signal to noise ratio of the recorded signal u_{Rx} is improved by averaging on 128 measurements. This permits to decrease the statistical errors [73]. The data are acquired for a time duration of 50 ns, recovering 2001 points per measurement.

The transmit and the receive antennas are two identical ridged horn antennas (Model 6100), as already used for the measurements in chapter 6. Their frequency domain behavior (transfer function) has been shown in Fig. 6.16. The considered configuration is a quasi-monostatic case.

The measurements have been taken in an anechoic chamber in order to minimize the reflections from the surrounding. A target object has been placed in the anechoic chamber at a distance of 2.9 m from the two antennas at the same height and in the main radiation direction.

For each target object 4 measurements have been performed: 2 co- and 2 cross-polarization measurements. In the calibration procedure time-gating has been applied for eliminating the antenna coupling and the background reflections. Moreover, for comparison, also the "empty room" has been measured (4 measurements: 2 for the co- and 2 for the cross-polarization).

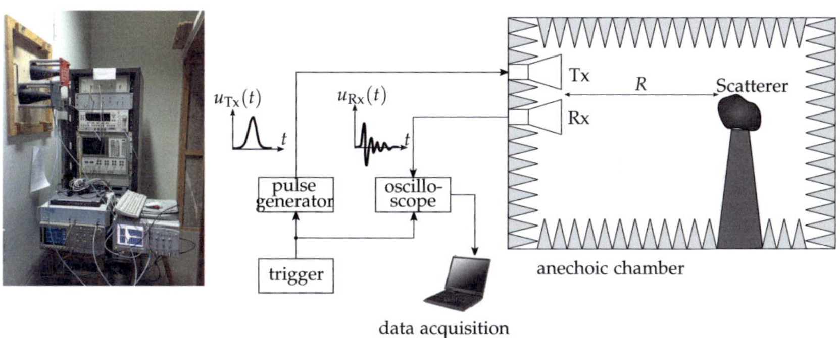

Figure 7.5: Measurement scenario for verification in the time domain.

7.4.1 Calibration Targets

In the following, the selected calibration targets and their measurement results are reported.

Flat Plate: The first selected calibration target is a flat plate with dimensions 17.5×17.5 cm^2. The measured plate impulse responses for the four different polarization components are shown in Fig. 7.6, after the deconvolution of the influence of the antennas, of the connecting cables and of the transmit pulse's derivative. The deconvolution has been performed using the Wiener deconvolution method, which is described in the Appendix, where also the choice of this method has been explained.

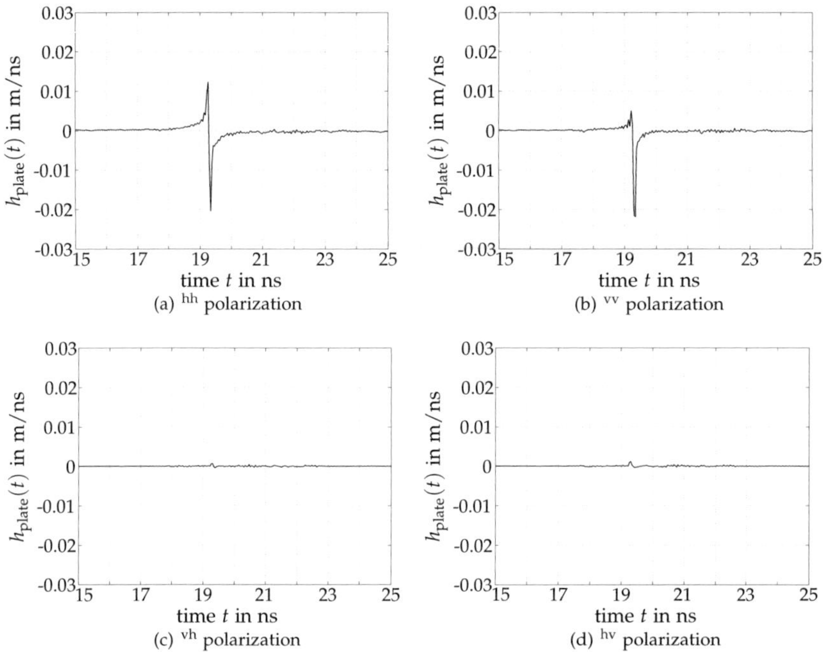

Figure 7.6: Measured $h_{Sc}(t)$ for the flat plate for the different polarization components.

Dihedral corner reflector, offset $0°$**:** The second calibration target is a dihedral corner reflector with $0°$ offset from the vertical. It has been positioned with the incident ray path perpendicular to its fold line. The dimensions of the used dihedral (which is symmetrical with respect to the fold line) are 17.5×17.5 cm^2. In Fig. 7.7 the measured plate impulse responses for the four different polarization components are shown.

Dihedral corner reflector, offset $45°$**:** The same dihedral corner reflector as above has been positioned with an offset of $45°$ with respect to the vertical. The measured plate impulse responses for the four different polarization components are shown in Fig. 7.8.

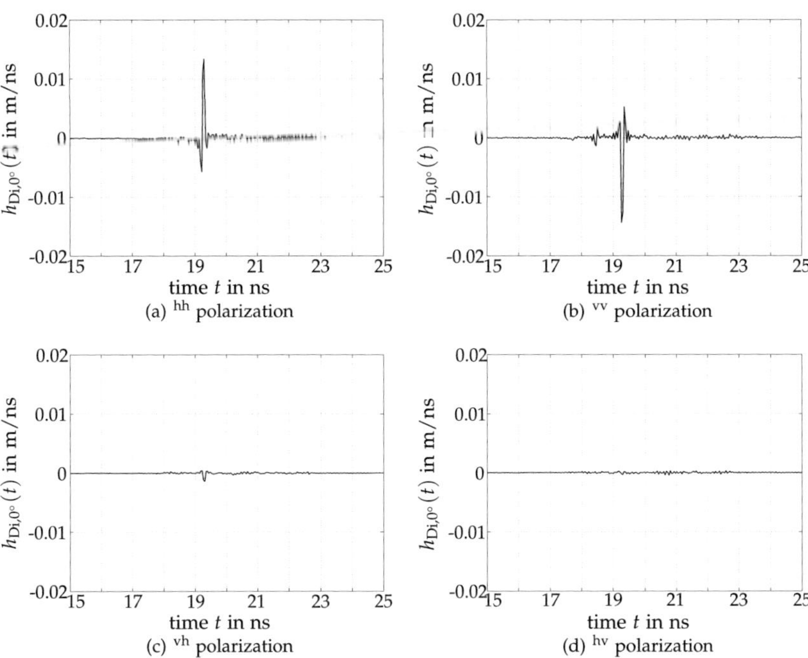

Figure 7.7: Measured $h_{\text{Sc}}(t)$ for the vertical dihedral with $0°$ offset from the vertical for the different polarization components.

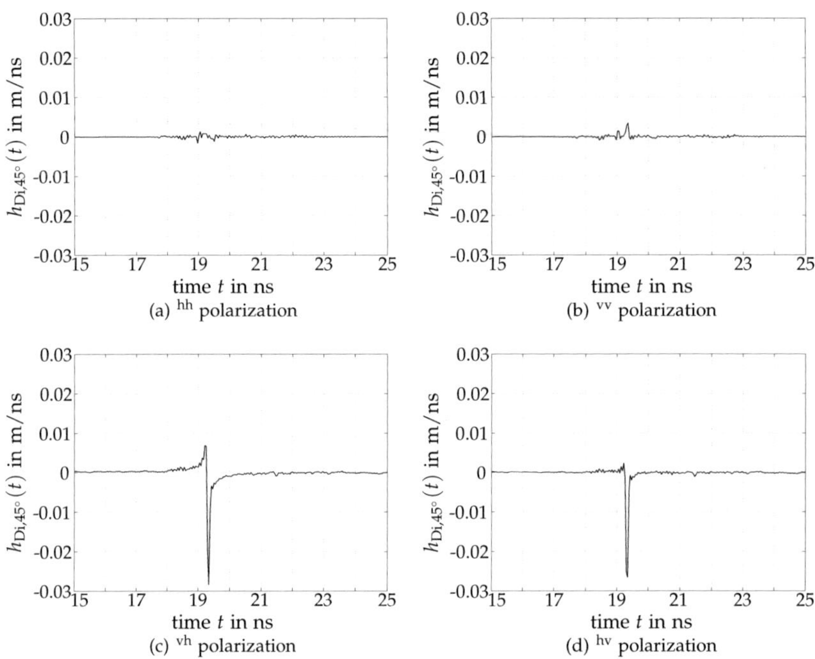

Figure 7.8: Measured $h_{Sc}(t)$ for the vertical dihedral with $45°$ offset from the vertical for the different polarization components.

7.4.2 Example of Calibration

The calibration procedure described in section 7.3 has been applied to the measurement results of the scattering from a conducting cylinder (hight 18.5 cm, diameter 6 cm), vertically positioned, in order to investigate the possible improvement of the measurement results after the calibration has been performed. The target has been positioned in the anechoic chamber at the same distance as the reference targets. The measured co-polarized (^{vv}) and cross-polarized (^{vh}) impulse responses before and after the calibration procedure are plotted in Fig. 7.9 and 7.10, respectively. The peak arrives after 19.4 ns, which corresponds to a distance of $2r = 5.8$ m and hence $r = 2.9$ m. As it can be seen from comparison of Fig. 7.10(a) and 7.10(b), the calibration permits to drastically decrease the cross-polarized measured impulse response, which would be zero for the cylinder.

A database with the impulse responses (for each polarization component) of different targets (sphere with 30 cm diameter, flat plate, vertical dihedral 0°, vertical dihedral 45° offset and vertical cylinder) has been constructed. Comparing, for each polarization, the measured calibrated impulse response of the target with the database, as explained in section 7.2.3, the results for the correlation and the peak have been calculated. From this comparison, it is possible to conclude that the measured object is the cylinder.

7.5 Conclusion

In this chapter the UWB Radar link in the time domain has been fully polarimetric mathematically described introducing a time domain version of the error-cube, previously defined only in the frequency domain. It has been seen that, in order to correctly determine the polarization components of the impulse response of a target, a calibration procedure is necessary. In the chapter a novel calibration procedure for UWB Radar has been introduced. This procedure is entirely performed in the time domain. Performing all the calibration operations in the time domain has significant advantages with respect to classical frequency domain methods. First of all, for each calibration target, only 4 measurements are necessary (2 for the co- and 2 for the cross-polarization) instead of the hundreds of steps at different frequencies in the frequency domain. Moreover, the calculation of the "empty room" is not necessary anymore, since time-gating can be performed, exploiting the different time of arrival of the background reflections, of the antenna coupling and of the signal backscattered from the target. By an example measurement with a cylinder it has been shown that the proposed calibration method significantly improves the measurement results. It has also be shown that the UWB impulse Radar calibration in the time domain also allows to perform polarimetric UWB target classification directly in the time domain.

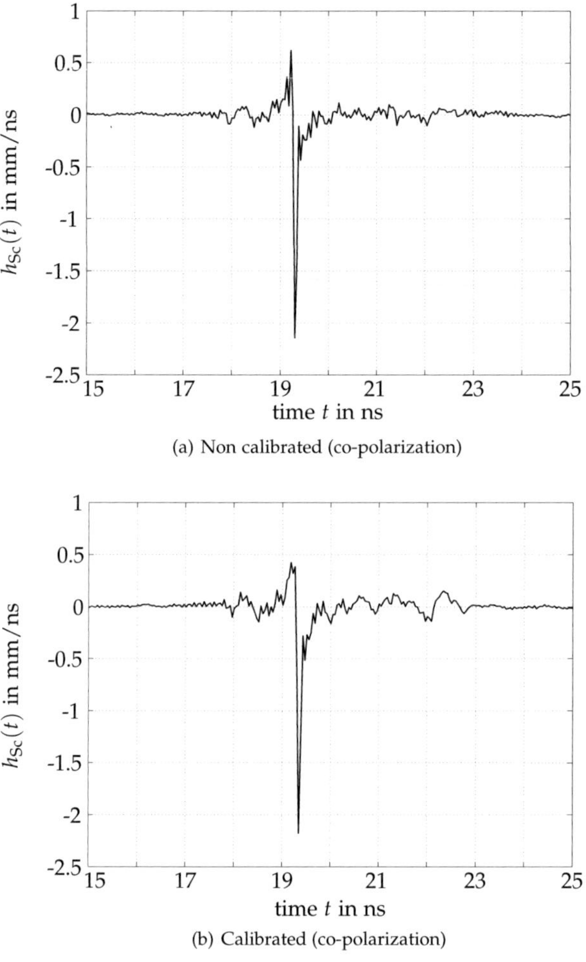

(a) Non calibrated (co-polarization)

(b) Calibrated (co-polarization)

Figure 7.9: Measured $h_{Sc}(t)$ for the cylinder before (a) and after (b) the calibration, co-polarization.

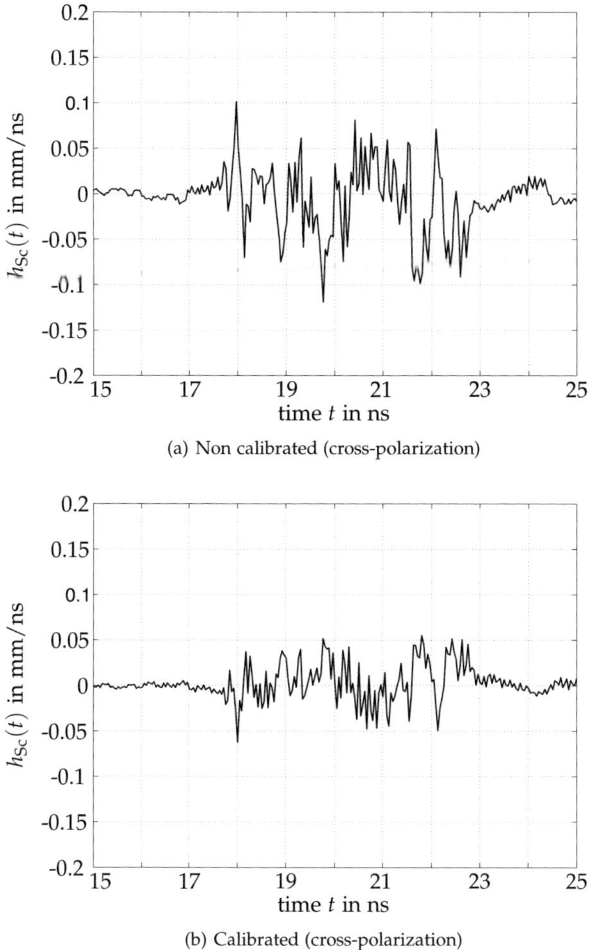

(a) Non calibrated (cross-polarization)

(b) Calibrated (cross-polarization)

Figure 7.10: Measured $h_{\mathrm{Sc}}(t)$ for the cylinder before (a) and after (b) the calibration, cross-polarization.

8 Conclusions

Ultra Wideband technology, thanks to its huge bandwidth, offers several advantages with respect to classical narrowband wireless technology, among them the transmission of high data rate, high resolution for Radar and localization, etc. Because of its advantages, in recent times many devices (filters, antennas, ...) for UWB applications have been developed. Consequently, a complete analysis not only of the UWB devices themselves taken separately, but also an investigation of the UWB radio and Radar link is necessary, in order to assess and optimize the performance of a complete UWB system.

In this thesis a performance study composed of two different stages has been performed. In the first stage the single UWB components (filters, antennas integrated with filters) have been separately investigated, in order to assess their time domain behavior and dispersiveness. In the second stage the single elements, previously separately investigated, have been analyzed from a system point of view, performing an analysis of the UWB radio link and of the UWB Radar link.

UWB devices are usually described through an investigation, which is conducted in the frequency domain. However, this analysis does not permit to directly assess the distortion and the dispersiveness caused by the device on the transmitted signal. Hence, in this thesis, a complete analysis of UWB components (such as filters, antennas, and the integration of filters and antennas) has been performed, which takes into account not only the frequency domain behavior but also conducts an investigation in the time domain, which permits to have a clearer insight into the distortion and the dispersiveness in the time domain due to the non-ideal behavior of the device itself. Together with the time domain analysis of the different components, also a correlation analysis has been introduced, in order to quantify the amount of distortion introduced on the signal by the device. These two analyses complement each other and permit to describe the time domain behavior of the component and its influence on the signal through few parameters. Differently from the frequency domain characterization, where the commonly used parameters are frequency dependent, these time domain parameters are single numbers, which have a direct relation to the pulse preserving capability and dispersiveness of the device. The mathematical background of these two analysis methods has been given in chapter 2.

The first UWB components investigated with both the time domain analysis and the correlation analysis in the time domain have been UWB filters (ref. to chapter 3). Using a statistical approach, it has been possible to quantify the impact of the filter frequency domain non-idealities (non-flat filter frequency mask and non-constant filter group delay time) on the time domain behavior. It has been seen that there exist bounds on the standard deviations of the passband filter mask and of the filter group delay time, which, if respected, permit to have only negligible distortion of the output signal with respect to the case of an ideal filter. In such a case, the decrement of the peak value and the increment of the ringing duration and of the full width at half maximum of the impulse response of a filter are negligible with respect to the impulse response of an ideal filter. Moreover, in order to assess the particular characteristics of different hardware realizations of UWB filters, different filter typologies (such as resonator stub filters, CPW filters, etc.) have been investigated both in the frequency domain and in the time domain. From the obtained results it has been seen that resonator stubs filters are affected from energy storage due to their resonant structure, which worsens their behavior (long ringing duration, high FWHM, low peak value) with respect to structures realized without resonator stubs. It has been seen that using an integration of lowpass and highpass sections based on microstrip-to-slotline transitions permits to decrease the occupied dimensions but suffers from the imperfect matching of the transition and of the reduced highpass behavior of the transitions.

After UWB filters have been investigated, they have also been integrated into antennas and the behavior of such an integrated device has been analyzed, both in the frequency domain and in the time domain (ref. to chapter 4). It has been seen that two different kinds of integration between antennas and filters are possible: the filter can be integrated in the antenna radiating element or in the antenna ground plane/feeding line. From simulation results and through measurements from fabricated prototypes it has been observed that the integration of a filter in the antenna ground plane/feeding line has a better performance with respect to the integration in the antenna radiating element.

Once the single components have been separately investigated, also an analysis firstly of the UWB radio link (ref. to chapter 5) and secondly of the UWB Radar link (ref. to chapter 7) has been performed from a system point of view. In the UWB radio system link analysis the signal at the receiver has been regarded in order to find criteria that permit to assess the quality of the UWB link from the perspective of the receiver including the non-ideal system behavior (taking into account in particular the non-idealities of the antennas). This approach has lead to the definition of two quality criteria which permit to quantify the worsening of the signal to noise ratio of a UWB radio link due to the non-ideal behavior of the antennas. These criteria allow for the determination of the spatial region where an antenna radiates strong signals, which are lowly distorted with respect to the signal radiated in the main beam direction. This has been done by evaluating the SNR at the receiver side in the case of a correlation receiver. It has been found out that the SNR is directly influenced by the angular-dependent behavior of the antennas. Hence, it has been seen that in order to quantify the variation of the SNR due to the antenna angular dependence, it is necessary to have a measure for quantifying this angular dependence. Through the analysis of the fidelity properties of UWB antennas it has been possible to determine the distortion that the transmitted signal suffers in the different angular directions with respect to the main beam direction in the time domain due to the non-ideal antenna behavior. Moreover, it has been seen that, in order to guarantee high SNR at the receiver, also the antenna impulse response has to present an high peak value, which has resulted in the definition of a joint criterion including both fidelity and peak value. The developed criteria have been investigated for two typical UWB antennas, resulting in the observation that both antennas provide differently shaped spatial areas of considerable size with good radiation characteristics.

Once the characterization of the non-idealities of the system has been investigated, strategies have been introduced in order to compensate for the spectral non-idealities of the single UWB components (antennas, pulse generator, etc.), (ref. to chapter 6). It has been seen that an optimization of the system performance, in terms of matching and fulfillment of the required mask, can be obtained by introducing a particular filter structure, which has both to select the required frequency interval and to pre-distort the signal in order to compensate for the spectral non-idealities. The required "shaping filter" transfer function has been mathematically derived and different prototypes have been developed, fabricated and tested. From measurement results it has been proven that with the usage of this particular filter design approach the transmit signal fits optimally to the given mask.

Together with the UWB radio link also the UWB Radar link has been analyzed from a system point of view, with a description entirely conducted in the time domain including a fully polarimetric measurement calibration procedure. It has been seen that a calibration in the UWB case is easier conducted in the time domain instead of in the frequency domain. In that case, only few measurement operations are required (2 co- and 2 cross-polarization measurements), compared to the usually hundreds of measurement steps in the frequency domain. Moreover, it has been found out that the "empty room" calibration is not necessary anymore since time gating can be easily applied. The calibration of an example measurement result has proven the operability of the proposed time domain calibration procedure and shown that it allows for drastically improving the results.

Concluding, this thesis has presented a complete investigation of both single UWB components as well as the UWB radio link and the UWB Radar link in the time domain. From the performed analysis it has been possible to assess the impact on the time domain behavior of the non-ideal characteristics of UWB devices both from a single device point of view and from a system point of view. Many quality criteria have been introduced to quantify the non-ideality of the single UWB devices and of the complete system in the time domain. The introduced criteria are easy to measure and have been verified through measurement results from prototypes. Also correction strategies have been introduced to compensate for the spectral non-idealities of UWB components. With all these strategies the system performance of Ultra Wideband systems can be drastically improved.

A Appendix

A.1 Practical recovery of time domain data from frequency domain data

Typical measurement equipment (e.g. vector network analyzer) operates in the frequency domain. The procedure for obtaining the impulse response includes the transformation of the data measured in the frequency domain to the time domain. Since it is only possible to recover samples of signals with a discrete frequency resolution, the time domain data are obtained via the Inverse Discrete Fourier Transform (IDFT) [81], [82], [83].

Let f_L and f_H be the start and stop frequencies of the measured frequency interval and N the number of points which are recovered, hence the measured bandwidth is $B = f_H - f_L$ and the frequency domain resolution is $\Delta f = B/N$. With a vector network analyzer a discrete version of the transfer function is measured, which is a sampled version with frequency resolution Δf. Hence, the impulse response can be recovered via IDFT. In this procedure the frequency range from 0 to f_L has to be filled with zeros. Firstly, the analytic transfer function $H^+(f)$ has to be constructed, according to eq. (2.4) and then applying to it the IDFT. The discrete impulse response results in [45]

$$
h(k\Delta t) = \Re \left[\sum_{n=0}^{\tilde{N}-1} H^+(n\Delta f) \cdot \exp\left[j\frac{2\pi}{\tilde{N}}kn \right] \right]
\tag{A.1}
$$

where $\tilde{N}$ is the number of points with the zero padding in the frequency range $0, \ldots, f_L$. The zero padding results in an interpolation. The time domain resolution is given by the measured bandwidth $\Delta t = 1/B$. In order to increase the interpolation and the time domain accuracy, an additional zero padding can be added at the end of the measured data in the frequency domain.

A.2 Practical recovery of frequency domain data from time domain data

In oder to obtain the frequency domain data from data $s(t)$ acquired in the time domain the Discrete Fourier Transform (DFT) is used. In order to correctly transform the data in the frequency domain, firstly the analytic signal $s^+(t)$ has been calculated applying the Hilbert Transform $\mathcal{H}$, namely [45]

$$
s^+(t) = s(t) + \mathcal{H}[s(t)] = s(t) + j \int_{-\infty}^{+\infty} s(t') \frac{1}{\pi(t-t')} dt'
\tag{A.2}
$$

This permits to obtain the complex time domain data starting from the measured data, which are real. Moreover, the respective Fourier transform is non zero only at positive frequencies. Then, the DFT is applied. Letting Δt be the time spacing of the acquired data and N the number of recorded points, the respective sampling frequency is $F_s = 1/\Delta t$ and the frequency spacing is $\Delta f = F_s/N$. Hence, in the frequency domain, the obtained signal is

$$S(k\Delta f) = \frac{1}{\Delta t} \sum_{n=0}^{N-1} s^+(n\Delta t) \cdot \exp\left[-j\frac{2\pi}{N}kn\right] . \tag{A.3}$$

What is obtained, is the analytic signal $S^+(f)$ in the frequency domain, which is related to actual signal spectrum as

$$S^+(f) = \begin{cases} 2S(f) & f > 0 \\ S(f) & f = 0 \\ 0 & f < 0 \end{cases} \tag{A.4}$$

The signal spectrum is hence obtained as $S(f) = S^+(f)/2$, for $f = k\Delta f$.

A.3 Model for a random Gaussian process

A random Gaussian process Ξ is characterized by three parameters [84], [85]

mean value μ it represents the mean value of the i-th realization ξ_i of the process;

standard deviation σ it represents the standard deviation of the i-th realization ξ_i of the process;

correlation separation Δ_{corr} it indicates the rate of variation of the process.

In Fig. A.1 three different realizations of a random Gaussian profile with the parameters $\mu = 0$, $\sigma = 1$ and different values of Δ_{corr} are illustrated.

A.3.1 On the correlation frequency separation

Different values of the parameter Δf_{corr}, which gives the rate of the variation of the process with frequency, have been investigated for the EU UWB case.

A lower value of Δf_{corr} means that the rate of variation of the ripple in the passband filter mask and of the GDT is higher. In that case, from simulations it has be seen that the influence of the GDT remains approximately the same for all investigated parameters.

On the other hand, the ringing is slightly influenced by a smaller Δf_{corr} in the ripple of the passband filter mask. The effect of a smaller Δf_{corr} is to lower the curves in Fig. 3.6-3.8 with respect to the vertical axis, i.e. a lower level of σ_A results in a higher level of ripple with respect to a higher value of Δf_{corr}. However, this variation is small compared to the selected value $\Delta f_{\text{corr}} = 0.2$ GHz.

A higher value of Δf_{corr} means that the rate of the variation of the ripples is lower. In that case the graphs show large areas where a quasi-ideal filter behavior is preserved, but this is not observed from measurement results. This because a higher Δf_{corr} means that in the frequency interval of interest the GDT is almost constant and the filter mask nearly flat, as it is possible to see in Fig. A.2, where various ripples are plotted for $\Delta f_{\text{corr}} = 0.5$ GHz.

From these observations and from measurement results it can be concluded that a good approximation of the rate of variation of the ripples in the passband mask and GDT for the EU UWB case can be obtained setting $\Delta f_{\text{corr}} = 0.2$ GHz.

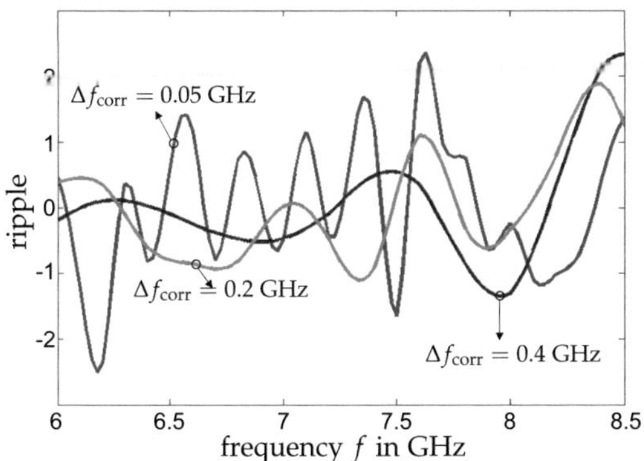

Figure A.1: Different realizations of a Gaussian process with $\mu = 0$, $\sigma = 1$ and various values of Δf_{corr}.

A.4 Deconvolution of time domain data

Letting $y(t) = h(t) * x(t)$, the simplest way to obtain $x(t)$ with the knowledge of $y(t)$ and $h(t)$ is applying the Fourier transform identity

$$y(t) = h(t) * x(t) \quad \circ\!\!-\!\!\!-\!\!\bullet \quad Y(f) = H(f) \cdot X(f) \tag{A.5}$$

and then dividing $Y(f)$ by $H(f)$, namely

$$x(t) = y(t) *^{-1} h(t) \quad \circ\!\!-\!\!\!-\!\!\bullet \quad X(f) = \frac{Y(f)}{H(f)} \, . \tag{A.6}$$

However, since measured data are affected by noise, at frequencies with low signal to noise ratio the noise level is increased through the division operation and hence the obtained $\tilde{x}(t)$ is different from $x(t)$.

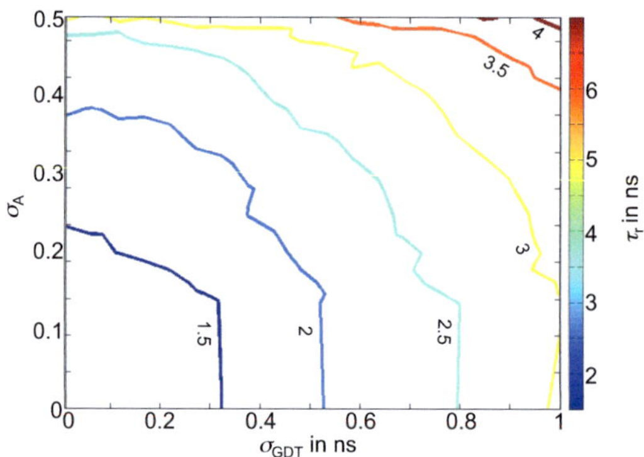

Figure A.2: Calculated ringing value τ_r for $r = 0.1$ and $\Delta f_{\mathrm{corr}} = 0.5$ GHz.

In literature different methods for solving the deconvolution problem are described [86], [87]. Authors have proposed methods based on the singular value decomposition (SVD) matrix factorization for obtaining Radar data in the time domain. Although this method has good performance, the drawback is the time required for the calculation of the matrices of the SVD when the number of the acquired data increases. Other authors have proposed to use a lowpass filter to reduce the noise level at high frequencies, where the division operation increases the noise level itself. The drawback of this method is the definition of the filter parameters.

In this thesis, a particular method has been used to deconvolve time domain data, the Wiener deconvolution.

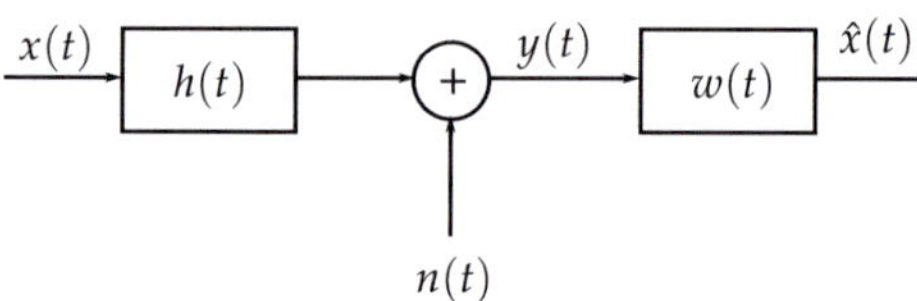

Figure A.3: Application of the Wiener deconvolution for obtaining an estimate of the input signal of a linear system

A.4.1 Wiener deconvolution

Given the linear system model shown in Fig. A.3, the Wiener deconvolution method permits to obtain an estimate $\hat{x}(t)$ of the unknown imput signal $x(t)$ with the knowledge of the system output $y(t)$, of the system impulse response $h(t)$ and of the noise level [88]. The Wiener deconvolution method permits to minimize the mean square error between the estimate $\hat{x}(t)$ and the signal $x(t)$. The method introduces a function $w(t)$, the definition of which in the frequency domain is [88]

$$W(f) = \frac{H(f)^* X(f)}{|H(f)|^2 X(f) + N(f)} = \frac{1}{H(f)} \frac{|H(f)|^2}{|H(f)|^2 + \underbrace{N(f)/S(f)}_{1/SNR(f)}} \qquad \text{(A.7)}$$

where $H(f)$ is the Fourier transform of $h(t)$, $X(f)$ is the spectrum of $x(t)$ and $N(f)$ the spectrum of the noise $n(t)$. The second equality has been obtained dividing both nominator and denominator by $X(f)$. The term $SNR(f) = X(f)/N(f)$ is the signal to noise ratio of the measurement.

Hence, the estimate of $x(t)$ is then obtained as

$$\hat{x}(t) = w(t) * y(t) . \qquad \text{(A.8)}$$

For practical deconvolution processing of the measurement results the SNR has been estimated through the knowledge of the noise of the instrumentation and the power of $y(t)$.

Bibliography

[1] C.E.Shannon, "A Mathematical Theory of Communications", *The Bell System Technical Journal*, no. 27, 1948.

[2] *Revision of Part 15 of the Commission's Rules Regarding Ultra-Wideband Transmission System*, ET-Docket 98-153, FCC, Feb. 14, 2002.

[3] *Official Journal of the European Union*, 23.2.2007, L.55/33-36.

[4] E.Pancera, G.Adamiuk, T.Zwick, W.Wiesbeck, "UWB out-of-phase network feeding a 2-element impulse radiating array", *Proceedings of the IEEE Antennas and Propagation Society International Symposium 2008*, San Diego, CA, USA, Jul. 2008.

[5] E.Pancera, W.Wiesbeck, "Differentially Fed Array for UWB Radar Applications", *3rd European Conference on Antennas and Propagation, EuCAP 2009*, Berlin, Germany, Mar. 2009.

[6] W.Soergel, *Characterisierung von Antennen fuer die Ultra-Wideband-Technik*, Dissertation, *Forschungsberichte aus dem Institut fuer Hoechstfrequenztechnik und Elektronik der Universitaet Karlsruhe (TH)*, Band 51, 2007.

[7] E.G.Farr, C.E.Baum, "Extending the Definitions of Antenna Gain and Radiation Pattern Into the Time Domain", *Sensor and Simulation Notes*, Note 350, Nov. 1992.

[8] E.G.Farr, C.E.Baum, C.J.Buchenauer, "Impulse Radiating Antennas, Part I", *Ultra Wideband, Short Pulse Electromagnetic*, vol. 2, pp.159-170, L.Carin, L.B.Felsen ed., Plenum Press, NY, 1995.

[9] J.S.McLean, H.Foltz, R.Sutton, "Pattern Descriptor for UWB Antennas", *IEEE Trans. on Antennas and Propagation*, vol.53, no.1, Jan. 2005.

[10] J.S.McLean, R.Sutton, H.Foltz, "The Effect of Source Pulse Shape on the Energy and Correlation Patterns of UWB Antennas", *European Conference of Wireless Technology 2004*, Amsterdam, 2004.

[11] M.Chamchoy, S.Promwong, P.Tangtisanon, J.-I.Takada, "Spatial Correlation Properties of Multiantenna UWB Systems for In-Home Scenarios", *ISCIT 2004*, Sapporo, Japan, Oct. 2004.

[12] J.S.McLean, H.Foltz, R.Sutton, "Differential Time Delay Patterns for UWB Antennas", *IEEE Int. Conf. on Ultra-Wideband, ICUWB08*, Hannover, Sep. 2008.

[13] E.Pancera, J.Timmermann, G.Adamiuk, T.Zwick, W.Wiesbeck, "Spatial distortion of the pulse transmitted by UWB antennas", *COST 2100* TD(08)521, Trondheim, Norway, Jun. 2008.

[14] G.Matthaei, L.Young, E.M.T.Jones, *Microwave Filters, Impedance-Matching Networks, and Coupling Structures*, Artech House, Norwood, MA, 1980.

[15] J.-S.Hong, M.J.Lancaster, *Microstrip Filters for RF Microwave Applications*, John Wiley & Sons, Inc. NY, 2001.

[16] J.-S.Hong, H.Shaman, "An Optimum Ultra-Wideband Microstrip Filter", *Microwave and Optical Technology Letters*, pp. 230-233, vol. 47, no.3, Nov. 2005.

[17] H.Shaman, J.-S.Hong, "A Novel Ultra-Wideband (UWB) Bandpass Filter (BPF) With Pairs of Transmission Zeroes", *IEEE Microwave and Wireless Components Letters*, pp. 121-123, vol. 17, no.2, Feb. 2007.

[18] W.-T.Wong, Y.-S.Lin, C.-H.Wang, C.H.Chen, "Highly Selective Microstrip Bandpass Filters for Ultra-Wideband (UWB) Applications", *APMC2005 Proceeding*, vol.5, Dec. 2005.

[19] K.Li, D.Kurita, T.Matsui, "Dual-Band Ultra-Wideband Bandpass Filter", *IEEE MTT-S International Symposium*, Jun. 2006.

[20] P.Cai, Z.Ma, X.Guan, X.Yang, Y.Kobayashi, T.Anada, G.Hagiwara, "A Compact UWB Bandpass Filter Using Two-Section Open-Circuited Stubs to Realize Transmission Zeros", *APMC2005 Proceedings*, 2005.

[21] A.Q.Liu, A.B.Yu, Q.W.Zhang, "Broad-Band Band-Pass and Band-Stop Filters with Sharp Cut-off Frequencies Based on Series CPW Stubs", *IEEE MTT-S International Symposium*, Jun. 2006.

[22] J.Sor, Y.Qian, T.Itoh, "Miniature Low-Loss CPW Periodic Structures for Filter Applications", *IEEE Transactions on Microwave Theory and Techniques*, pp. 2336-2341, vol. 49, no.12, Dec. 2001.

[23] J.Gao, L.Zhu, W.Menzel, F.Boegelsack, "Short-Circuited CPW Multiple-Mode Resonator for Ultra-Wideband (UWB) Bandpass Filter", *IEEE Microwave and Wireless Components Letters*, pp. 104-106, vol. 16, no.3, Mar. 2006.

[24] N.-W.Chen, K.-Z.Fang, "An Ultra-Broadband Coplanar-Waveguide Bandpass Filter With Sharp Skirt Selectivity", *IEEE Microwave and Wireless Components Letters*, pp. 124-126, vol. 17, no.2, Feb. 2007.

[25] L.-Zhu, H.Shi, W.Menzel, "Coupling Behaviors of Quarter-Wavelength Impedance Transformers for Wideband CPW Bandpass Filters", *IEEE Microwave and Wireless Components Letters*, pp. 13-15, vol. 15, no.1, Jan. 2005.

[26] A.Zverev, *Handbook of Filter Synthesis*, John Wiley and Sons, Inc., N.Y., 1967.

[27] W.Zschunke, "Bestimmung guter Ausgangsnaeherungen fuer die Optimierung von Filtern zur Datenuebertragung und Impulsformung", *NTG- Fachberichte* no. 51, pp. 69-76, VDE-Verlag 1975.

[28] W.Zschunke, *Beitrag zur Formung und Entzerrung von Empfangsfunktionen bei der Datenuebertragung*, Universitaet Stuttgart, Habilitationsschrift, 1972.

[29] M.I.Sobhy, E.A.Hosny, "Microwave Filter Design in the Time Domain", *Microwave Symposium Digest, MTT-S International*, vol. 81, no. 1, Jun. 1981.

[30] E.Pancera, C.Sturm, T.Zwick, W.Wiesbeck, "Impact of the UWB Filter Frequency Domain Non-Idealities on the Filter Transient Behavior", *Frequenz, Journal of RF-Engineering and Telecommunications*, pp. 164-170, vol. 62, no.7-8, Jul./Aug. 2008.

[31] E.Pancera, J.Timmermann, T.Zwick, W.Wiesbeck, "Signal Optimization in Ultra Wideband Systems", *GeMiC09*, Munich, Germany, Mar. 2009.

[32] J.Timmermann, A.A.Rashidi, P.Walk, E.Pancera, T.Zwick, "Application of Optimal Pulse Design in Non-ideal Ultra-Wideband Transmission", *GeMiC09*, Munich, Germany, Mar. 2009.

[33] B.Parr, B.Cho, K.Wallace, Z.Ding, "A Novel Ultra-Wideband Pulse Design Algorithm", *IEEE Comm. Lett.*, vol. 7, no. 5, May 2003.

[34] L.Yang, G.Giannakis, "Ultra-Wideband Communications, An Idea Whose Time Has Come", *IEEE Signal Processing Magazine*, vol. 21, no. 6, Nov. 2004.

[35] E.Pancera, D.Modotto, A.Locatelli, F.M.Pigozzo, C. De Angelis, "Novel Design of UWB Antenna with Band-Notch Capability", *European Microwave Week 2007, ECWT*, Munich, Oct. 2007.

[36] S.W. Su, K.L. Wong, and F.S. Chang, "Compact Printed Ultra-Wideband Slot Antenna with a Band-Notched Operation", *Microwave Opt. Technol. Lett.*, vol. 45, no. 2, pp. 128-130, Apr. 2005.

[37] E.Pancera, C.Sturm, G.Adamiuk, T.Zwick, W.Wiesbeck, "Electromagnetic Analysis of Band-Notched UWB Antennas", *GeMiC08*, Hamburg-Harburg, Mar. 2008.

[38] E.Pancera, J.Timmermann, T.Zwick, W.Wiesbeck, "Time Domain Analysis of Band Notch UWB Antennas", *EuCAP 2009*, Berlin, Mar. 2009.

[39] E.Farr, C.Baum, C.J.Buchenauer, "Impulse Radiating Antennas, Part II", *Ultra-Wideband, Short Pulse Electromagnetics, 2*, L.Carin, L.B.Felsen ed., Plenum Press, New York, 1995.

[40] S.R.Cloude, P.D.Smith, A.Milne, D.M.Parkes, K.Trafford, "Analysis of the Time Domain Ultrawideband Radar Signals", *Ultra-Wideband, Short Pulse Electromagnetics*, H.Bertoni ed., Plenum Press, pp.445-456, 1993.

[41] G.C.Gaunaurd, H.C.Strifors, S.Abrahamsson, B.Brusmark, "Scattering of Short EM-Pulses by Simple and Complex Targets Using Impulse Radar", *Ultra-Wideband, Short Pulse Electromagnetics*, H.Bertoni ed., Plenum Press, pp.437-444, 1993.

[42] W.A.van Cappellen, R.V.de Jongh, L.P.Ligthart, "Potentials of Ultra-Short-Pulse Time-Domain Scattering Measurements", *IEEE Antennas and Propagation Magazine*, vol. 42, no. 4, Aug. 2000.

[43] E.Pancera, A.Bhattacharya, E.Veshi, T.Zwick, W.Wiesbeck, "Ultra Wideband Impulse Radar Calibration", *IET International Radar Conference 2009*, Guilin, China, Apr., 2009.

[44] D.M.Pozar, *Microwave Engineering* 2nd ed., John Wiley and Sons, Inc, 1998.

[45] J.G.Proakis, *Digital Communications*, 3rd ed., McGraw-Hill, New York, NY, USA, 1995.

[46] W.Soergel, W.Wiesbeck, "Influence of the Antennas on the Ultra-Wideband Transmission", *EURASIP Journal on Applied Signal Processing*, vol.3, pp.296-305, 2005.

[47] D.Lamensdorf, L.Susman, "Baseband-pulse-antenna techniques", *IEEE Antennas and Propagation Magazine*, vol.36, no.1, Feb. 1994.

[48] P.A.Meyer, "Ueber Filter mit angenaehert rechteckfoermiger Impulsantwort", *NTZ*, vol. 18, no. 5, pp. 249-255, 1965.

[49] Y.-S. Lin, W.-C. Ku, C.-H. Wang, C.H. Chen, "Wideband Coplanar-Waveguide Bandpass Filters with Good Stopband Rejection", *IEEE Microwave Wireless Component Letters*, vol. 14, no. 9, Sep. 2004.

[50] E.Pancera, T.Zwick, W.Wiesbeck, "Compact UWB Bandpass Filter based on Microstrip-to -Stripline Transitions", *Journal of the European Microwave Association*, pp. 104-106, vol. 4, no.4, Dec. 2008.

[51] H.Wang, L.Zhu, "Ultra-Wideband bandpass filters using back-to-back microstrip-to-CPW transition structure", *Electronic Letters*, pp. 1337-1338, vol. 41, no.24, Nov. 2005.

[52] N.Thomson, J.-S.Hong, "Compact Ultra-Wideband Microstrip/Coplanar Waveguide Bandpass Filter", *IEEE Microwave and Wireless Components Letters*, pp. 184-186, vol. 17, no.3, Mar. 2007.

[53] B.Schueppert, "Microstrip/Slotline Transitions: Modeling and Experimental Investigation", *IEEE Trans. Microwave Theory Tech.*, vol. 36, no. 8, Aug. 1988.

[54] N.I.Dib, L.P.B.Katehi, G.E.Ponchak, R.N.Simons, "Theoretical and Experimental Characterization of Coplanar Waveguide Discontinuities for Filter Applications", *IEEE Transactions on Microwave Theory and Techniques*, pp. 873-882, vol. 39, no.5, May 1991.

[55] N.I.Dib, G.E.Ponchak, L.P.B.Katehi, "Theoretical and Experimental Study of Coplanar Waveguide Shunt Stubs", *IEEE Transactions on Microwave Theory and Techniques*, pp. 38-44, vol. 41, no.1, Jan. 1993.

[56] R.N.Simons, *Coplanar Waveguide Circuits, Components, and Systems*, John Wiley & Sons, Inc. NY, 2001.

[57] K.Hettak, N.Dib, A.Omar, G.-Y. Delisle, M.Stubbs, "A Useful New Class of Miniature CPW Shunt Stubs and its Impact on Millimeter-Wave Integrated Circuits", *IEEE Transactions on Microwave Theory and Techniques*, pp. 2340-2349, vol. 47, no.12, Dec. 1999.

[58] K.Hettak, N.Dib,A.-F. Sheta, S.Toutain, "A Class of Novel Uniplanar Series and Their Implementation in Original Applications", *IEEE Transactions on Microwave Theory and Techniques*, pp. 1270-1275, vol. 46, no.9, Sep. 1998.

[59] *Revision of Part 18 of the FCC-Industrial, Scientific and Medical Equipment*, CFR Title 47 Part 18, FCC, Oct. 1, 2007.

[60] *IEEE Standard Definitions of Terms for Antennas*, IEEE Std, 145-1993, SH16279, Mar. 1993.

[61] S.H. Lee, J.K. Park, and J.N. Lee, "A Novel CPW-Fed Ultra-Wideband Antenna Design", *Microwave Opt. Technol. Lett.*, vol. 44, no. 5, Mar. 2005.

[62] W.S. Lee, W.G. Lim, and J.W. Yu, "Multiple Band-Notched Planar Monopole Antenna for Multiband Wireless Systems", *IEEE Microwave Wireless Compon. Lett.*, vol. 15, no. 9, pp. 576-578, Sep. 2005.

[63] W.C. Liu, P.C. Kao, "CPW-FED Triangular Antenna with a Frequency-Band Notch Function for Ultra-Wideband Applications", *Microwave Opt. Technol. Lett.*, vol. 48, no. 6, Jun. 2006.

[64] S.W. Qu, J.L. Li, Q. Xue, "A Band-Notched Ultrawideband Printed Monopole Antenna", *IEEE Ant. Wireless Propag. Lett.*, vol. 5, 2006.

[65] M.J.Toro, "Phase and Amplitude Distortion in Linear Networks", *Proc. of the IRE*, pp. 24-36, Jan. 1948.

[66] G.L. Turin, "An Introduction to Matched filters", *IRE Transactions on Information Theory*, vol. 6, no.3, pp.311-329, Jun. 1960.

[67] D.-H.Kwon, "Effect of Antenna Gain and Group Delay Variations on Pulse-Preserving Capabilities of Ultrawideband Antennas", *IEEE Trans. on Antennas and Propagation*, vol.54, no.8, pp.2208-2215, Aug. 2006.

[68] D.M.Shan, Z.N.Chen, X.H.Wu, "Signal Optimization for UWB Radio Systems", *IEEE Trans. on Antennas and Propagation*, vol.53, no.7, Jul. 2005.

[69] E.Pancera, C.Sturm, J.Fortuny-Guasch, T.Zwick, W.Wiesbeck, "Full 3D spatial time domain analysis of UWB antennas", *Proceedings of the IEEE Antennas and Propagation Society International Symposium 2008*, San Diego, CA, USA, Jul. 2008.

[70] Y.Wu, A.Molisch, S.-Y.Kung, J.Zhang, "Impulse Radio Pulse Shaping for Ultra Wideband (UWB) Systems", *IEEE Symposuim on Personal, Indoor and Mobile Radio Communication Proceedings*, 2003.

[71] X.Wu, Z.Tian, T.Davidson, G.Giannakis, "Optimal Waveform Design for UWB Radios", *IEEE Trans. on Signal Processing*, vol. 54, no. 6, Jun. 2006.

[72] Z.N.Chen, X.H.Wu, H.F.Li, N.Yang, M.Y.W.Chia, "Considerations for Source Pulses and Antennas in UWB Radio Systems", *IEEE Trans. on Antennas and Propagation*, vol.52, no.7, pp.1739-1748, Jul. 2004.

[73] W.Wiesbeck, S.Riegger, "A Complete Error Model for Free Space Polarimetric Measurements", *IEEE Trans. on Antennas and Propagations*, vol.39, no.8, pp. 1105-1111, Aug. 1991.

[74] W.Wiesbeck, D.Kaehny, "Single reference, three target calibration and error correction for monostatic, polarimetric free space measurements", *Proceedings of the IEEE*, vol.79, no.10, pp. 1551-1558, Oct. 1991.

[75] W.Wiesbeck, *Radar System Engineering*, Lecture Script, IHE, Universitaet Karlsruhe (TH).

[76] H.Mott, *Antennas for Radar and Communications, a Polarimetric Approach*, J. Wiley and Sons, New York, 1992.

[77] Y.Chevalier, Y.Imbs, B.Beillard, J.Andrieu, M.Jouvet, B.Jecko, M.Le Goff, E.Legros, "UWB Measurements of Canonical Targets and RCS Determination", *Ultra-Wideband, Short Pulse Electromagnetic*, vol. 4, pp. 329-334, Heyman et al. ed., Kluwer Academic/Plenum Publisher, NY, 1999.

[78] E.Pancera, T.Zwick, W.Wiesbeck, "Correlation Properties of UWB Radar Target Impulse Responses", *IEEE Radar Conference 2009, RadarCon09*, Pasadena, CA, USA, May, 2009.

[79] G.T.Ruck, ed., *Radar Cross Section Handbook*, Plenum Press, New York, 1970.

[80] *Picosecond Pulse Labs*, Model 3600 Impulse Generator Instruction Manual.

[81] W.Soergel, F.Pivit, W.Wiesbeck, "Comparison of Frequency Domain and Time Domain Measurement Procedures for Ultra Wideband Antennas", *Proceedings Antenna and Measurement Techniques Association, 25th Annual Meeting and Symposium*, pp. 72-76, Irvine, CA, Oct. 2003.

[82] D.Iverson, "Extracting real samples from complex sampled data", *IEEE Electronics Letters*, vol. 27, no. 21, pp. 1976-1978, 1991.

[83] S.L.Marple Jr., "Computing the discrete-time "analytic" signal via FFT", *IEEE Transactions on Signal Processing*, vol. 47, no. 9, pp. 2600-2603, 1999.

[84] A.Papoulis, *Probability, Random Variable and Stochastic Processes*, McGraw-Hill, NY, 2002.

[85] J.A.Ogilvy, *Theory of Wave Scattering From Random Rough Surfaces*, IOP Publishing, Philadelphia, PA, 1992.

[86] J.Rahman, T.K.Sarkar, "Deconvolution and Total Least Squares in Finding the Impulse Response of an Electromagnetic System from Measured Data", *IEEE Trans. on Antennas and Propagation*, vol. 43, no. 4, Apr. 1995.

[87] E.J.Rothwell, W.Sun, "Time Domain Deconvolution of Transient Radar Data", *IEEE Trans. on Antennas and Propagation*, vol.38, no.4, Apr. 1990.

[88] A.Papoulis, *Signal Analysis*, McGraw-Hill, NY, 1977.

Curriculum Vitae

Personal Information

Name	Elena Pancera
Birthplace	Verona, Italy
Date of Birth	July 13, 1981

School Education

2000	Maturità Scientifica at Liceo Scientifico Statale "A. Messedaglia" (High School), Verona, Italy

Studies and Research

2000 - 2005	Telecommunication Engineering at the University of Padova, Italy
2005	Laurea summa cum laude in Telecommunication Engineering at the Faculty of Engineering, University of Padova, Italy
2005 - 2007	Scientific collaboration for a PRIN Project on Ultra Wideband at the Department of Information Engineering, University of Padova, Italy
since March 2007	Research Assistant (Wissenschaftliche Mitarbeiterin) at the Institut für Hochfrequenztechnik und Elektronik (former Institut für Höchstfrequenztechnik und Elektronik) Universität Karlsruhe (TH) Research Topics: Ultra Wideband Antennas, Filters, Radar, Microwave components Teaching Activities: Hochfrequenztechnik, Hochfrequenzlaboratorium Teaching for courses organized by the "European School of Antennas"